# ANIMAL ECOLOGY TO-DAY

# MONOGRAPHIAE BIOLOGICAE

**EDITORES**

F. S. BODENHEIMER  W. W. WEISBACH
Jerusalem  Den Haag

VOL. VI

UITGEVERIJ DR. W. JUNK -- DEN HAAG -- 1958

# ANIMAL ECOLOGY TO-DAY

BY

F. S. BODENHEIMER
Professor Emeritus of Zoology
Hebrew University of Jerusalem

UITGEVERIJ DR. W. JUNK – DEN HAAG – 1958

*Dedicated*

*to*

*four great pioneers of animal ecology*

VICTOR SHELFORD – Urbana

RICHARD HESSE – Berlin

A. J. NICHOLSON – Canberra

P. A. ERRINGTON – Ames

Copyright 1958 by Uitgeverij Dr. W. Junk, Den Haag
Printed in the Netherlands
Zuid-Nederlandsche Drukkerij N.V. – 's-Hertogenbosch

CONTENTS

Chapter                                                                          page

I. Physiological and Ecological Life-Tables . . . . . . . . 12
    Life-intensity and longevity, ecological and physiological life-expectation, age structure of populations, ecological ages

II. The Life-History and its Ecological Interpretation   40
    Physical ecology, life cycles of some lady-beetles, climograms, bonitation of *Ceratitis*, sense ecology and behaviour

III. First Approach to an Analysis of some Animal Populations . . . . . . . . . . . . . . . . . . . . . . . . . . 67
    *Drosophila*, aphids, *Schistocerca*, epidemiology of malaria, voles in Israel, muskrats in Iowa

IV. Animal Populations in Equilibrium . . . . . . . . . . 112
    The case for biotic equilibrium and for weather control, the theorems of VOLTERRA and of NICHOLSON, climatic control, biotic balance, biological control, equilibrium

V. Is the Animal Community a Dynamic or merely a Descriptive Conception? . . . . . . . . . . . . . . 164
    Some definitions, animal communities in sea and on land, superorganism or empirical combination

VI. The Interaction of Environment and of Heredity within the Organism . . . . . . . . . . . . . . . . . . . . 202
    Heredity, population genetics, diapause, sex-determination, geographical variation, adaptation

VII. Why Human Ecology . . . . . . . . . . . . . . . . . 237
    Human fertility, environmentalism in sociology and geography, erosion and similar effects of human activity in forestry, sea fishery, and husbandry

# INTRODUCTION

"Everyone who writes a textbook on any branch of experimental science must set down as many wrong statements as right; he cannot carry out most experiments himself, he must rely on the testimony of others and often take probability for truth. Thus a compendium is a monument of the time when the facts were collected and it must be renewed and rewritten again and again. But while fresh discoveries are accepted and a few chapters improved, others perpetuate misleading experiments and erroneous deductions."

from *The Theory of Colour*, by GOETHE.

A colleague[1] stated recently in an ecological analysis that he had studied some parts of the population problem intensively and that therefore his findings were based on definite facts; others had been studied by him relatively little, so that on these he offered suggestive ideas based on insufficient evidence; and still others he had barely studied so that he gave in these cases unbased theories. This is a good and sincere statement which should serve to introduce all treatises on the theory of animal populations.

Despite the apparently great diversity of opinions in modern population theory, the contrasting views stem mainly from very similar situations. Every one of us who has devoted a lifetime of research on animal populations, has by necessity, because of the limitations of the human mind, proceeded in one way. We all studied one aspect thoroughly, found this special approach fertile, and continued to follow this promising path. This limitation of perspective however has blinded us to greater or lesser degree to other colours or perspectives of nature's picture. We have all, at least during early development, overemphasized one aspect and neglected many others. Once this difficulty is fully recognised, the time has come for an attempt to synthesize all the serious work in our field without any false compromise. We have endeavoured to give here a tentative synthesis, lacking perfection as it may still be, in the hope that we have nonetheless demonstrated that it is both possible and feasible. All different approaches can be united in the study of a population as a whole, thus merely complementing one another.

The situation is similar to that described for philosophy by A. J. AYER[2], who stated: "I maintain that there is nothing in the nature of philosophy to warrant the existence of conflicting philosophical "schools". And I attempt to substantiate this by providing a definite solution of the problems which have been the chief sources of controversy between philosophers in the past."

Many ways have been ventured by nature to solve the problem of "natural balance", and it is for the common good that almost every path has been trodden, often to the end, with eager stubbornness. Every one of these attempts has caught one corner of the truth, without which we would be unable to comprehend the whole.

This book is not written against anyone or against any theory. Everybody has the right to build up his philosophy according to his own experience and suiting his own temperament. We accord due respect to every idea of a serious worker, and have permitted ourselves merely that same liberty to build our own philosophy based on our life experiences, principally in the acquisition of facts and principles of animal ecology, as revealed by others as well as ourselves. It is not our intention to convince the reader or produce a "sole truth" of animal ecology. Nature moves in many directions and makes use of all possible roads to the solution of a problem. A good illustration of this point is the climbing of a tree-trunk–spirally by a woodpecker, straight by a squirrel. We present the views of an individual who tried hard to rid himself of prejudices, to learn what seemed to him to be important. The synthesis gained may be of interest to others still seeking their way through the embarrassing multiplicity of natural phenomena.

This book is not sophisticated. Its purpose is to be a "common sense" book, with the background of prolonged experience. This combination should be the aim of every young worker in animal ecology. Even if his predilections lead him from autecology to some kind of "biocoenotics" as the basis of his ecological research, he should neglect neither common sense nor experience. In our science, as in every science "All roads lead to Rome" as long as these fundamentals are safeguarded.

Between the "Problems of Animal Ecology" (1938, now long out of print) and the present "Animal Ecology To-day" lies a world of experience – ecological facts, the shape of the problems and a maturation in the author's experience. Thus the "Animal Ecology To-day" is practically a new book. We consider the two greatest steps in ecological progress, both connected with the names of A. J. NICHOLSON and P. A. ERRINGTON, to be:

(1) The first puts numeric changes of population in the centre of all research; and

(2) The second is the principle of compensation.

These steps have altered our outlook on animal ecology more than anything else. We may here add another principle, now developing: The difference between homology and analogy must be much more stressed than hitherto. We often considered casual, accidental, non-structural parallels and coincidences as causally significant, whereas they actually are merely cases of analogy.

As stated above, the principal aim of ecological research is to explore the fluctuation of populations and their (homologous) causes; that of the environment is never an aim in itself, only that of the population. A factor causing high mortality in early youth may be numerically insignificant and have no influence on the size of the reproducing population, as compensatory factors will probably replace most or all of this early mortality at later stages. We were thus certainly inaccurate in stating in the "Problems" that all factors are destructive in direct proportion to the percentage per stage destroyed by each. The population flow is much less dependent on any single factor or group of factors at any stage of development, as any part of the population killed by one factor is interchangeable and replaceable by many other factors (density-dependent as well as density-independent ones). The important population base remains the number of reproducing and breeding adults at the season of reproduction, where such a season exists. All losses of the other ages are of little importance as they are either pre- or post-reproductive. Thus, we may distinguish between two regular fluctuations within every population in time. The one is the regular and steady decimation of the pre- and post-reproductive ages in every generation, but the real fluctuations which count are the annual changes in numbers of the population in the reproductive age from one reproducing season to the next one.

Ever-increasing importance is ascribed to the organism itself. True, it is an inseparable part of the ecoworld, meaning that it is an integral part of its own (so-called) environment, reacting to it in behaviour and physiology, but never a passive object to the environmental changes. It also influences its own "environment" as by its reactions it actively participates in all actions, re-actions and co-actions of its ecoworld.

In the meantime the problems of "synecology" become more and more formal descriptions which are useful schemata of the human mind but which scarcely enter the problems of dynamic ecology, so far mainly restricted to intraspecific population problems.

We are now as remote from any decisive conclusion as before, but the continuous rearrangement of our thought leads to what we believe an ever clearer comprehension of the problems of animal ecology. It is dubious whether more could be achieved in principle even in the future of ecology or in that of any other science[3]. We believe the real solutions of any problem, exceeding mere description and its coordination, are beyond the grasp of the human mind.

Many problems analysed in this book were new twenty-five years ago, when the "Problems" were conceived. We hope that this book takes full account of the advances of ecology during this period. The main problems treated are: The contrast between physiological and ecological longevity still seems one of the best introductions to the problems of animal ecology. The correlations between longevity and life-intensity offer important suggestions from the borderland of physiology and ecology.

The ecological interpretation of the life-history of the individual has passed its preliminary stages and becomes an increasingly valuable instrument of prognosis. The analysis of population growth has been studied in the "Problems" by means of *Drosophila*. Here we give the story of the long road of errors in the theory of density as a decisive population factor, interesting in that as much can be learned from errors as from truth. Some sections analyse the ecological lifehistory of a few other common animals such as *Schistocerca gregaria*, aphids, the epidemiology of malaria, of *Microtus guentheri*, outbreaks of the desert locust and of muskrats in Iowa.

The problem of so-called biological equilibrium, whose solution has been attempted from many different angles, is now ripe for real synthesis doing justice to them all, provided that nonambiguous definitions are given of basic concepts.

Animal communities are generally treated in the same sense as classical plant sociology treats its associations. Our analysis clearly shows that the belief in a supraorganismic integration of life-communities is based on intuition – which may be true or not, but is certainly not proven by induction. They remain, in our regions at least, valuable statistical units for description and somehow even for classification. The interaction of environment and heredity is one of the central problems of general biology. We here try to demonstrate the need for close cooperation of genetics and ecology in population dynamics, diapause, sex-determination, geographical variation, and adaptation. It is most improbable that any one of these phenomena may be understood alone, without realising the background of the other. We wish to stress here that phenotype and genotype are not the expression of heredity vs. environment, but that both have a genetical background which is manifested on different ecological backgrounds. It would be misunderstanding the author's intention were this chapter to be taken as a strict support of the Lamarckian theory. All experimental proof is on the side of Mendelism, but all experience together is still far from being conclusive to the exclusion of any other solution. We merely plead for an unprejudiced approach to the problem, but we wish to make it clear that the writer believes that Lamarckism is one of the solutions of evolution, that we personally do not believe in any kind of superorganismic biocoenotics etc. We think everyone has the right to his own concept of truth. Negation of this freedom of opinion means negation of science.

We have added a chapter, "Why Human Ecology?" We intended originally to deal with that problem in a special book but we doubt whether our life-span would suffice for this task. The implications of consistent neglect of our own ecological problems may involve us in future catastrophes far greater than those menacing us in the shape of the H-bomb. These problems should be to-day very much more in the centre of ecological discussion. The human future is at stake. We have omitted here all the problems connected with the spiritual changes of man and mankind as we do not believe that man will change in this respect in any near future.

It would have been simple to paraphrase the entire text in our own words, but we have preferred to let the authors of important theses and observations speak in their own words.

This book is not easy reading. Some documentation is always appended in the form of tables or of histograms. It was an erroneous assumption on the part of the author in the "Problems", that these tables or histograms could prove any opinion. All opinions are formed by intuition, always strictly conditioned by the writer's prejudices. Each result usually leaves the way open for at least two contradictory conclusions. Reader and author can only further development by keeping their minds open for any present or future contradiction between fact and theory. It is the privilege of the scientist not to wish to be right, but to wish to further the discovery of truth, if it be in agreement with or in contradiction to his own earlier opinion. The selection of our examples is, of course, not at random, but careful and without bias, as far as possible. Omission of these documents might have rendered the text easier, but easy reading should still be avoided in the present stage of animal ecology, where the formulation of theories is so simple and the basis of facts so narrow. We have replaced a number of illustrations, published in the "Problems" by others representing the present trend of animal ecology more effectively.

Young as our science is, we never forget that the mere accumulation of facts is not science – their interpretation and synthesis are. In recent years there has been perhaps too much loose thinking; mere speculation has been developed, certainly no desirable development. It is only long and rich personal experience in nature that gives the ecologist the background for fruitful thought. A. J. NICHOLSON is perhaps a splendid illustration: – The speculative and deductive thoughts of his youth, almost a generation ago, are now growing into rich fruition, backed by a wealth of observation and experiment. Another good illustration is P. A. ERRINGTON, who always had open eyes for work in nature and who there did find answers to many questions, steadily finding new ones.

# I
# PHYSIOLOGICAL AND ECOLOGICAL LIFE-TABLES AND CONNECTED PROBLEMS
(LIFE-INTENSITY, AGE-STRUCTURE OF ANIMAL POPULATIONS, RELATION BETWEEN THE ECOLOGICAL AGES)

## 1. Physiological Life-Expectation

Every analysis of life-expectation begins with a study and definition of physiological longevity. Physiological longevity may be defined as that life duration which a healthy individual may expect to live under optimal environmental conditions until dying by senescence. Such an individual, having exhausted its inborn vital potentialities, dies because death is the end of every organized life.

Any individual out of an animal population bred under optimal conditions or under an optimal sequence of conditions (weather, food, density, parasites, diseases, episites, etc., and accidents excluded) and selected from a thoroughly healthy and genetically homogeneous stock will die a physiological death, conditioned only by senescence. Assuming that conditions are always optimal and that genetically the stock is healthy and homogeneous, an ideal curve of longevity, or death, must be expected. All individuals should survive until senescence and death set in, within a short period following a sharp decline. The main difference would be due to the different longevity of the sexes and to the minor individual differences which are always met with, even in a genetically very similar stock. Ecological conditions induce a heavy infant mortality and a permanent but lower mortality until senescence, when the last individuals die.

However, no animal lives throughout life under optimal conditions, and the physiological life-expectation is not verified in nature. Also, no two individuals of the same population live under exactly the same conditions during their life, and no two individuals of gamogenic origin are exactly alike in their hereditary constitution, inborn resistance, and vitality *.Whatever is said here for individuals holds still more for populations of the same species.

Even in the absence of enemies and diseases, the following factors induce a deviation from the ideal death curve:

(A) Genetic: different genetic constitution; not thoroughly healthy constitutions.

(B) Ecological: deviation from optimal conditions during one, some, or all stages (concerning weather, food, density, enemies, etc.).

---
* See note a, p. 267

This is the reason why even those laboratory breedings which approach the ideal death-curve closest show deviations from the beginning onward.

Our actual knowledge of physiological longevity is rather limited[1]. FLOWER[2] has compiled all observations available for vertebrates, including the experience gotten from zoological gardens. However, whereas there is some sense in studying the average longevity of birds and mammals in their natural habitat, and also the average longevity of long-lived poikilothermous animals like tortoises, crocodiles, lizards, fish, bee- or ant-queens, etc. under the same conditions, the conception of longevity loses any meaning as soon as we try to apply it to the smaller and shorter-lived poikilothermous animals.

But before discussing this complicated situation we must decide as to the limits of individual life. With regard to death, this is simple, and the physiological inquiry as to the exact moment of death is of little importance. It rarely happens that the experimenter or the observer is in doubt as to whether an animal is dead or not. And even should there arise difficulties, means are easily found to settle any doubts.

It is rather astonishing how little thought has been given to the beginning of individual life. In mammals, the moment of birth, in birds, reptiles, and fish the moment of hatching from the egg has commonly been accepted as the beginning of life. With regard to insects and other invertebrates, all developmental stages, except the imago, have been compared with the embryological development in mammals, etc., and have therefore been excluded from any inquiry into longevity[3]. The entomologist will regard this statement as absurd. The homologies and analogies of developmental stages, however, will always offer opportunity for alternative or varying interpretation. The only possible way of settling these sources of misunderstanding is the definition: Longevity is measured as the period beginning with the egg-fertilization (or with the beginning of cleavage in parthenogenetically developing animals, or with the beginning of increased cell-division in animals reproducing by budding) and ending with the death of the individual. This solution, besides guarding us from embarrassing puzzles, ends the unjustified exclusion of the embryonic period from the life-period. Physiological as well as ecological and hereditary factors influence this important stage, as well as later ones, and the environmental resistance of this stage is often much less than that of any later age. Thus, for example, the mammalian embryo is doubtless physiologically the most sensitive of all ages. It is, however, extremely well-protected- mechanically, by the mother's belly, by its even temperature, moisture, osmotic pressure, i.e. by the steadiness of the so-called "milieu fixe interne" of CLAUDE BERNARD, creating an almost ideal homoeostatic environment. Ecological mortality in this most sensitive stage is consequently very low. Actually, the greatest ecological mortality occurs to the neonates, which, endangered by inability to retain their tissue-water, lose it rapidly and die. This is the case in human infant toxicosis, which may also be reduced considerably by change of environmental conditions and by medicaments.

## 2. Life-Intensity and Longevity

Speed of development, as well as adult longevity, depend on environmental factors, mainly temperature and air humidity. High temperature and optimal humidity shorten the life-span, low temperatures lengthen it in poikilothermous animals. Theoretically, at the threshold of development, animals could be kept in "cold-storage" for an indefinite period of time. Practically, they will die slowly after some (generally rather long) period has passed. Temperature (as well as other factors) is thus not only a factor affecting the quality of the environment, but it also influences the speed of life and life processes.

In principle it seems admissible that within certain temperature limits of tolerance, which are specifically different and sometimes even different with populations of the same species, the total life-intensity is more or less equal at all temperatures. This means that long-lived individuals display the same total activity at low temperatures, as measured by metabolic processes, etc. until their natural death as do the short-lived individuals at high temperatures, where intensity is much higher per unit of time.

Physiologists have promoted the theory that, within the same species, longevity – apart from genetic constitution – is inversely proportional to the intensity of life. This statement, first made in respect of mammals[4] (man excluded), maintained that they show the same total sum of basic metabolism after the completion of growth. It has not been confirmed by other students[5].

| Species | Longevity | Netto calories (Rubner)[4] | Brutto calories Puetter[5] |
|---|---|---|---|
| Man | 60 years | 725,800 | 725,000 |
| Horse | 30 years | 163,900 | .. |
| Cattle | 26 years | 141,090 | 253,000 |
| Dog | 9 years | 164,000 | .. |
| Cat | 8 years | 223,000 | .. |
| Guinea-pig | 6 years | 265,000 | .. |
| Lion | 31 years | .. | 323,000 |
| Elephant | > 100 years | .. | > 1,660,000 |
| Camel | > 30 years | .. | > 1,210,000 |

With regard to poikilothermous animals, however, this rule seems to hold (within certain limits of temperature, etc., genetic constitution being equal).

Pearl[6] has given a stimulating demonstration, in this respect, on the growth and longevity of cantaloupe seedlings nourished only by the food contained in their cotyledons, showing inherited vitality. Putting the mean of the experiment at 100%, the result was as follows:

|  | Seedlings which lived | | |
|---|---|---|---|
|  | 14 | 15 | 16 days |
|  | % | % | % |
| Total duration of life | 95 | 102 | 109 |
| $CO_2$ rate per day of growth during growth-period only | 104 | 102 | 81 |
| Dry matter metabolized | 105 | 100 | 87 |
| Water metabolized | 104 | 100 | 88 |

In *Drosophila* similar results have been obtained. In series of flies reared at 18° and 28° C the following results were obtained as regards relative differences in size:

| Character | Difference in size between 18° and 28° C flies expressed as percentage of the size of 18° C flies | |
|---|---|---|
|  | Males | Females |
| Femur length | 7.0 | 5.7 |
| Tibia length | 6.9 | 6.0 |
| Wing length | 13.4 | 10.0 |
| Wing breadth | 13.1 | 9.4 |
| Mean of four characters | 10.1 | 7.8 |

Similarly, differences occur in the longevity of flies bred at these temperatures:

| Temperature during adult life | Difference in longevity between 18° and 28° flies expressed as percentage of the longevity of 18° C flies | |
|---|---|---|
|  | Males | Females |
| 18° C | 19.5 | 7.6 |
| 25° C | 16.1 | 12.7 |
| 28° C |  | 7.0 |
| Mean | 17.8 | 10.2 |

"These results indicate, with a considerable degree of probability, that in these experiments the quantitative effects of temperature differences upon the biological processes concerned in growth are of approximately the same order of magnitude as the quantitative effects of temperature differences upon the biological processes concerned in the determination of duration of imaginal life. This is the kind of numerical result which would be expected on the rate of living theory of life duration, because in both cases the effect of increased temperature is to speed up the rate of the biological processes involved. In the 18° C flies we have a slow rate of energy expenditure in growth and during imaginal life (flies very inactive), and we should therefore expect on the theory the lengthened duration of imaginal life which we observe. In the 28° C flies there is a short developmental period and a consequent rapid rate of energy expenditure during growth and during imaginal life (flies very active). This leads to the expectation of a short duration of imaginal life, which is in fact observed."[7]

In *Daphnia magna*[8] the product of heart-beats per second (life-average) and average longevity is constant for both sexes:

$$♂\ 4\cdot 3 \text{ beats} \times 37\cdot 8 \text{ days} = 162\cdot 5$$
$$♀\ 3\cdot 7 \text{ beats} \times 43\cdot 8 \text{ days} = 162\cdot 1$$

whereas the duration of life of females exceeds that of males by 14·6%. Longevity as a function of metabolic rate is also in agreement with experiments on influence of temperature. The influence of temperature on the intensity of physiological processes is well known, but it seems that temperature plays no part in the coefficient of utilization of energy in the living system, nor in the coefficient of utilization of nutritive substances for growth. This has been shown for *Proteus vulgaris*[9], for muscle activity[10], for the amount of sugar used up to form 1 g of dry substance in *Aspergillus niger*, which is the same for different temperatures[11]. In tadpoles the amount of substance stored during growth, as compared with that of substance burned, is practically independent of temperature[12]:

| Temperature, °C | 8 | 10 | 14 | 21 |
|---|---|---|---|---|
| Age at which gills disappear, days | 30 | 22 | 20 | 8 |
| Coefficient of nitrogen utilization | 0.75 | 0.73 | 0.75 | 0.75 |

The same is true for the gaseous metabolism of insect pupae. In *Tenebrio molitor*[13] the total quantity of $CO_2$ produced during the pupal stage was:

| Temperature, °C | 18.8 | 20.9 | 23.7 | 27.3 | 32.3 |
|---|---|---|---|---|---|
| $CO_2$, l/kg | 59.0 | 59.6 | 59.1 | 58.0 | 59.3 |

A surprising result was obtained by PARKER[14]. In breeding *Melanoplus atlanis* at different temperatures, the total amount of food consumed (dry weight) by 10 nymphs was as follows:

| Temperature, °C | 22 | 27 | 32 | 37 |
|---|---|---|---|---|
| Duration of nymphal stage, days | 94 | 54 | 27 | 25 |
| Food consumed, milligrams | 4,079 | 4,311 | 4,098 | 3,988 |

With alternating changing temperatures as well, no significant difference was observed in the total amount of food consumed during the nymphal stage. Similar results were obtained on caterpillars of *Phlyctaenia ferrugalis*[15].

The theory that rate of metabolism and longevity react inversely is greatly supported by the fact that insects, after having been treated with poisons or having been subjected to other unfavourable influences in sublethal dosages, show a shortness of life-duration parallel to the intensity of the treatment. Females of *Ephestia kühniella* show, after treatment with $CO_2$, at $17.5°$ C, a shortening of adult life[16]:

| Age at treatment in days | Exposure in seconds | How many days aged | Theoretical life-duration |
|---|---|---|---|
| 11.5 | 0 | 0 | 23.0 |
| 11.0 | 5.5 | 0.5 | 22.5 |
| 10.0 | 15.3 | 1.5 | 21.5 |
| 9.0 | 25.0 | 2.5 | 20.5 |
| 8.0 | 34.5 | 3.5 | 19.5 |
| 7.0 | 44.5 | 4.5 | 18.5 |
| 6.0 | 54.0 | 5.5 | 17.5 |
| 5.0 | 63.8 | 6.5 | 16.5 |
| 4.0 | 73.5 | 7.5 | 15.5 |
| 3.0 | 83.0 | 8.5 | 14.5 |
| 2.0 | 94.0 | 9.5 | 13.5 |
| 1.0 | 108.0 | 10.5 | 12.5 |
| 0.6 | 126.0 | 10.9 | 12.1 |
| 0.53 | 150.0 | 10.98 | 12.02 |

These results are confirmed empirically and controlled by the parallel measurements of the hysteresis reaction.

Most interesting results are obtained from observations on field-bees. The worker-bee, after a constant development period of 21 days and more or less constant 10 days' service as both a nurse-bee and as house-bee, turns into a field-bee of varying longevity. PHILLIPS[17], an outstanding authority on bee life, concludes his observations:

"The worker bees, which develop from eggs identical with those from which queens issue, live 6–10 weeks in summer and possibly

6 months or longer in winter. Those which emerge in time to take part in the gathering of a heavy honey crop usually live about 6 weeks, but if no nectar is available, the length of life is extended. Those workers which emerge at the close of summer are the ones which must live until the following spring if the colony is to survive, for there is no rearing of brood in normal colonies in winter. It is obvious that the length of life is influenced to a marked degree by the amount of work which they are called upon to do. Similarly queens live longer if they are not compelled to lay such large numbers of eggs. If bees winter badly, so that they are compelled to produce much heat, they often die, in the spring, faster than they are replaced by oncoming bees, a condition known as spring dwindling. Not all bees die in summer within 6 weeks of their emergence, for if all brood rearing is prevented, it may happen that the last bee will not die for 4 months. All these facts indicate that a bee is born with a definite amount of ability to do work and when its energy is expended the bee dies. It must not be concluded that bees have no recuperative power, but it is obvious that their term of life is limited by the amount of work they do".

This conception is confirmed quite independently by two other observations. LUNDIE[18] counted, by means of an electrical device, the number of bees leaving and returning to the hive during the season of activity. 3.16% of the bees did not return, i.e. died in the field. This would mean that one bee makes 31.65 trips before death overtakes it in the field. Whereas the number of deaths within the hive is rather limited, about 32 trips per worker-bee would be the average amount of work per field-bee. BODENHEIMER & BEN-NERYA[19] obtained a similar result by dividing the total number of trips per annum through the total production of bees during the same period:

$$6,533,880 : 220,145 = 29.68 \text{ trips per field-bee.}$$

The agreement between both results is rather close, especially if the difference in climate and methods is considered. The results confirm the theory that longevity of field-bees depends rather on the number of trips than on any other factor. The same is true of the total adult life. The amount of work done by the nurse- and house-bee is more or less constant during the active season. When no brood is to be nursed, the amount of energy normally spent on these activities is economized and – along with the stop of field trips in winter in colder climates – permits the survival of these bees throughout the interruption of the brood-rearing cycle.

All these facts tend to show that the lifespan to be expected is determined by specific and individual constitution, a given constant, and by intensity of life as measured by metabolism and activity, a varying magnitude. The life of a fly living at 30° C is shorter than that of one living at 20° C, but the life-intensity of both individuals is equal. And what is true of longevity is also true of adult life-span alone.

Finally, the temperature-sum rule is in itself a strong argument in this direction. This is not only correct for development, but also for partial physiological processes. Thus in fish the duration of the breeding season is inversely proportional to temperature[20].

The temperature-sum rule states that the product of time and effective temperature (= environmental temperature — development threshold) is constant. The hyperbola constructed in this way agrees with empirical data on development from the specific development threshold to an upper zone, above which heat does not accelerate development but actually becomes injurious, inhibiting its speed and finally leading up to death.

It is possible to construct not one but three such hyperbolae: the first one (A) lasts from the egg, which is very close to fertilization in insects, to first oviposition of the adults. The entire development and pre-oviposition periods are included in this period. Another hyperbola (C) may be constructed, based from the egg to average adult mortality. The difference between C and A gives the average adult longevity. From an epidemiological point of view, both of these hyperbolae are of little value. We may therefore construct an epidemiological average (B) at the period when half of the eggs of the hatching generation are laid.

We could base hyperbola C, as well, on the longest longevity at the corresponding temperature, and thus obtain the maximal life-span. One glance at such curves suffices to see that it is impossible to give absolute values for longevity of poikilothermous animals. Other conditions being constant, temperature is the dominating factor. However, quite a series of other factors, as population density, food-conditions, humidity, etc., interact in nature with temperature, not considering the constitutional, intra- and inter-population differences.

The only way out is to put the maximal longevity (or the average longevity) at 100, respectively, the half-time mortality at 50. Only in this way longevities at different temperatures, or at different habitats, may be compared. This necessity for comparison of different generations of poikilothermous animals or of corresponding generations in different habitats or countries on a common scale has been hitherto overlooked.

It must, however, be kept in mind that the entire conception of constant life-intensity is not a mathematical rule, even when neglecting the genetic background.

At high temperatures most insects of the same genetical stock are smaller (as indeed are mammals and birds) than at lower ones.

LESLIE SMITH[21] has explained some of the reasons, and probably the decisive ones, for *Hyalopterus pruni*. The rate of feeding of this aphid was related to temperature, and it was found that for a given increase in temperature the rate of feeding did not increase as rapidly as the rate of reproduction. This results in the relative starvation of young born at higher temperatures. Likewise, the smaller summer eggs of *Aonidiella aurantii*[22] cannot be expected to grow to the same size as winter adults, at least not at the same total expense of energy. However, the relations between the percentage variations in morphological differences and longevity are closely parallel in *Drosophila*[7].

Environmental changes, such as those of temperature, often lead to a physiological reconstruction of the organism concerned. Marine animals living in different latitudes often show a higher basic metabolism in the warmer climate[23].

Rates of Pulsation of the Dorsal Vessel of *Perinereis cultrifera (Polychaeta)* (after M. Fox 1939)

| Provenience | Month | Mean weight of individual | Temperature | Mean Pulsation for 10 minutes |
|---|---|---|---|---|
| Plymouth | X | 3 to 4 g. | 10° C | 64 |
|  |  |  | 14.5 | 92 |
|  |  |  | 20 | 132 |
| Tamaris | VIII/IX | 0.3 | 14.5 | 72 |
|  |  |  | 20 | 90 |
|  |  |  | 25 | 103 |

The pulsations of the blood vessel are thus adapted to the environmental temperatures in that the rate for animals living near Plymouth is the same at 14° as that of the animals from Southern France at 20°. At a given temperature the rate for the northern individuals is quicker than that of the southern ones, despite the tenfold greater size of the northern animals. Echinoderms, crustaceans, molluscs all behave similarly. This clearly indicates that the physiology of the same animal species in different environments is not the same – most important fact for the proper evaluation of the following chapter.

We find this important phenomenon in insects. A good illustration is the fact that the thermal sum needed for the total development of one generation is often much smaller in the centre or in the warmer parts of a distribution area than towards its northern, cooler border. This has been shown, e.g., in the Pine Shoot Moth *Evetria buoliana* on Mt. Carmel and on the mountains of Southern Sweden[23]. Rabbits adapted to constant heat are found to lose, to a considerable degree[24] their ability to produce heat and keep warm under chilling emergencies. Those adapted to a cooler environment are more able to meet chilling conditions by increased heat production, but succumb to excessive heat much more readily than do those adapted to constant warmth. A few hours' cooling counteracts the depressive effects of heat, and keeps the metabolism just as responsive to chilling as does a constantly cool environment. Animals living in a cold or changeable environment rapidly exhaust their glycogen supplies when chilled, while combustion is much slower in those from the hot room[25].

It is therefore not surprising that the results obtained are not all concordant. The statement of TITSCHAK[26] that the quantity of food eaten by the cloth-moth *Tineola biseliella* is much larger at 15° C than at 30° C, but that the relative consumption of food (per unit weight) is much higher at 30° C than at 15° C, needs further inquiry. Possibly one quality, e.g. water, was the limiting factor, and physiological food-utilization of the other substances of the food was incomplete.

However, the essential explanation is given by KOZHANTSCHIKOW[27]. In studying some physiological processes of *Blatta orientalis* he obtained the following zones:

| Process | Minimum | Maximum | Pessimum |
|---|---|---|---|
| | °C | °C | °C |
| Gaseous exchange | 0 | 40–45 | 45–55 |
| Irritability | 5–10 | 35–40 | 40–45 |
| Frequency of respiration | 10–15 | 40–45 | 45–55 |
| Heart-activity | 10–15 | 30–35 | 35–40 |
| Feeding | 10–15 | ? | 30–35 |

He characterizes physiological optimal conditions as the temperature zone in which all physiological processes proceed with moderate and comfortable rhythm. The speed of processes is not maximal, but may be accelerated to a higher maximum (more than doubled). These rhythms lead to a moderate relative effect of the activity of the individual processes and to their maximal total effect. The maximum of any process cannot be considered as the optimum.

The main point is that each of the different physiological processes has a lower threshold, optimum, maximum, and upper limit of its own, and these different processes are co-ordinated to a total and more or less efficient rhythm within the organism. A change in environmental conditions induces not only a change in the single processes but makes a re-coordination necessary. As long as such new co-ordinates are not forced into too large deviations from the optimal co-ordination, the length-intensity rule seems to hold good.

The great advantage gained by mammals and birds over all other animals is precisely the "milieu fixe interne" which renders them largely independent of changes and fluctuations in environmental conditions.

### 3. Ecological Life-Expectation

Optimal conditions, or a sequence of optimal conditions, to be maintained throughout the life of any individual in nature, and physiological death by senescence are practically wanting in nature. Unfavourable environment, heterogeneous genetical composition, diseases, enemies, accidents, etc., kill individuals of every natural population by scores, mostly during the most sensitive stages of development, before they have grown adolescent or mature. And for the large majority of the relatively few individuals which survive the full period of sexual maturity, the post-reproductive period is much shortened. This actual life-history of a population or a generation is its ecological life-history.

Fig. 1, IA illustrates the actual difference in the physiological and ecological life-history in the case of the desert locust in a large cage. In social animals (man, termites, ants, bees) the ecological mortality is often greatly reduced, and both curves differ only slightly. In some cases, as in that of the honey-bee, practically no mortality occurs before the bee has left the hive as field-bee. The analysis of the factors causing the difference between the curves of the physiological and the ecological lifetables is one of the main tasks of animal ecology.

A small series of life-tables based on breedings or, in the case of man, on comprehensive census work will be produced below. It should be kept in mind, however, that breedings, if carefully carried out, offer sub-optimal conditions to the animal, and are therefore apt to lower the death-rate by providing protection from enemies, diseases, and accidents, by providing suitable food, sufficient in quantity and quality, and by maintaining a comfortable temperature and humidity. We can, therefore, learn relatively little from these life-tables with regard to their ecological longevity. But by varying certain factors separately, we may study the influence of the individual major factors on longevity.

We shall begin with PEARL's work on *Drosophila*, as it analyses the influence of a series of external factors on longevity (Fig. 2). First we compare the vitality of wild *Drosophila* and some mutants of the second chromosome, with characters which have not before been connected with other than morphological changes of minor importance[7].

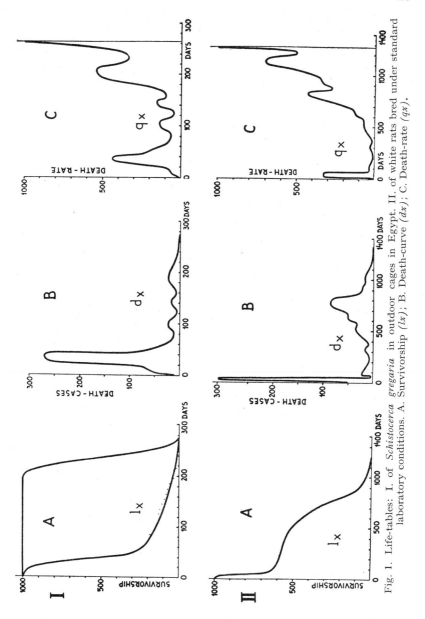

Fig. 1. Life-tables: I. of *Schistocerca gregaria* in outdoor cages in Egypt. II. of white rats bred under standard laboratory conditions. A. Survivorship $(lx)$; B. Death-curve $(dx)$; C. Death-rate $(qx)$.

This result with mutants of a highly homozygotic strain of wild *Drosophila* gives a clear idea as to the amount of constitutional variation in longevity to be expected in any heterozygotic insect population. In different strains of wild *Drosophila* adult longevity at constant conditions varies from 26 to 54 days[28, 29].

The influence of temperature[7] is illustrated by the following figures, representing adult longevity of *Drosophila* females, under different conditions:

| Temperature during adult life | Mean duration of life of flies reared at | |
|---|---|---|
| | 18° C | 28° C |
| 18° C | 70.61 | 65.25 days |
| 25° C | 40.96 | 35.75 days |
| 28° C | 30.67 | 28.52 days |

**Table I.**
Survivorship distribution of adults of wild *Drosophila* and single mutants at 25° C (both sexes included, not weighed). (After GONZALEZ)

| Age in days | Number of survivors up to indicated age | | | | | | |
|---|---|---|---|---|---|---|---|
| | Wild | Quintuple | Black | Purple | Vestigial | Arc | Speck |
| 1 | 1.000 | 1,000 | 1,000 | 1,000 | 1,000 | 1,000 | 1,000 |
| 4 | 998 | 955 | 991 | 995 | 996 | 986 | 985 |
| 7 | 992 | 683 | 987 | 939 | 938 | 938 | 977 |
| 10 | 971 | 424 | 974 | 873 | 809 | 873 | 963 |
| 13 | 948 | 269 | 970 | 791 | 664 | 813 | 932 |
| 16 | 901 | 168 | 949 | 711 | 527 | 760 | 900 |
| 19 | 882 | 115 | 928 | 622 | 394 | 709 | 884 |
| 22 | 858 | 75 | 890 | 558 | 306 | 643 | 855 |
| 25 | 831 | 48 | 854 | 495 | 206 | 575 | 827 |
| 28 | 799 | 31 | 832 | 425 | 150 | 464 | 792 |
| 31 | 743 | 13 | 768 | 338 | 105 | 402 | 745 |
| 34 | 705 | 6 | 690 | 218 | 62 | 316 | 687 |
| 37 | 592 | 5 | 619 | 151 | 43 | 241 | 643 |
| 40 | 519 | 1 | 503 | 99 | 30 | 165 | 614 |
| 43 | 424 | .. | 420 | 67 | 8 | 107 | 544 |
| 46 | 342 | .. | 371 | 34 | 2 | 62 | 468 |
| 49 | 247 | .. | 293 | 20 | .. | 32 | 417 |
| 52 | 194 | .. | 241 | 7 | .. | 15 | 333 |
| 55 | 117 | .. | 196 | 3 | .. | 2 | 273 |
| 58 | 88 | .. | 119 | 1 | .. | 1 | 207 |
| 61 | 67 | .. | 65 | .. | .. | 1 | 149 |
| 64 | 45 | .. | 30 | .. | .. | .. | 102 |
| 67 | 32 | .. | 20 | .. | .. | .. | 50 |
| 70 | 20 | .. | 14 | .. | .. | .. | 24 |
| 73 | 8 | .. | 3 | .. | .. | .. | 15 |
| 76 | 1 | .. | 1 | .. | .. | .. | 2 |
| 79 | .. | .. | .. | .. | .. | .. | 2 |
| 82 | .. | .. | .. | .. | .. | .. | .. |

The influence of population density on longevity in *Drosophila* is to be gathered from the following compilation[28]:

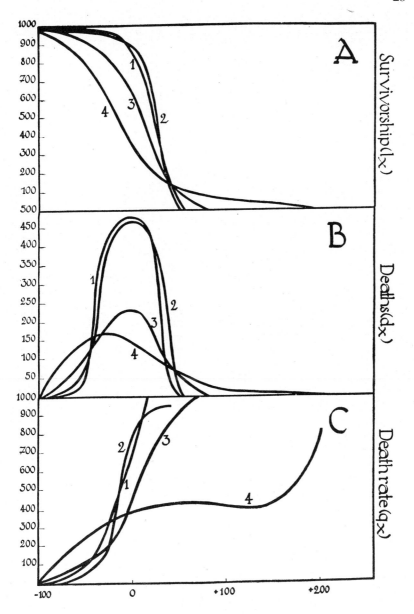

Fig. 2. Lifetable curves of *Drosophila*. A. Survivorship *(lx)*; B. Death curve *(dx)*; C. Death rate *(qx)*. The abscissa shows longevity, the mean longevity being assumed to be 0, all other ages being expressed as + or — percentage deviation from this mean. 1: Starved wild *Drosophila*; 2: Vestigial starved *Drosophila*; 3: Wild *Drosophila*; 4: Vestigial *Drosophila* (After PEARL).

| Initial density (flies) | | Average density (flies) | | Mean longevity (days) | |
|---|---|---|---|---|---|
| 2 | 55 | 1.8 | 44.7 | 27.3 | 40.0 |
| 4 | 75 | 3.3 | 59.7 | 29.3 | 32.3 |
| 6 | 95 | 5.0 | 74.5 | 34.5 | 27.2 |
| 8 | 105 | 6.7 | 80.4 | 34.2 | 24.2 |
| 10 | 125 | 8.2 | 94.4 | 36.2 | 19.6 |
| 15 | 150 | 12.4 | 111.9 | 37.9 | 16.2 |
| 25 | 200 | 20.7 | 144.5 | 37.5 | 11.9 |
| 35 | | 28.9 | | 39.4 | |

Fig. 2 illustrates the different important curves of the life-table for wild and vestigial *Drosophila* flies, starving and normal. It is clear that the mortality from egg to hatching of fly is excluded from these curves, which illustrate survivorship $(l_x)$ until the age referred to, the number of deaths $(d_x)$ occurring within each age-group of the population and death-rate $(q_x)$ of every age, provided the whole population is composed of that age-class only. For simplification's sake, all figures are referred to an initial population of 1,000 individuals born.

For futher information on the ecological life-history of *Drosophila* the reader is referred to Chapter III.

Breedings of the desert locust *(Schistocerca gregaria)* in Egypt[30] served as a basis for calculating the mortality curve during summer. The breedings were performed in large cages at moderate to suboptimal conditions. The biotic factors (enemies) were essentially excluded, but probably a few lizards entered the cage and fed on the locusts. The basic data were:

Stage . . . . . egg  larva I  larva II  larva III  larva IV  larva V  Total
Duration in days 27    8        9         11         8        12       75
Percentage mor-
  tality . . . . 16.5  35.5    12.5       8.4       6.1       1.8      80.8

Sexual maturity was reached in June/July on the 155th day of life, i.e. the pre-oviposition period lasted 80 days. The hyperbolae are as follows:

| | A<br>pre-oviposition period | B<br>at oviposition period | C<br>A-B until average death |
|---|---|---|---|
| Threshold (c), °C | 17.4 | 17.4 | 17.4 |
| Thermal constant (th.c.), days-degrees | 978 | 1,322 | 1,930 |
| Duration in days at: | | | |
| 35° C | 56 | 75 | 110 |
| 30  C | 77 | 104 | 153 |
| 25  C | 129 | 174 | 254 |
| 20  C | 376 | 508 | 743 |
| Average percentage of period | 50.5 | 17.5 | 32.0 |

Among the 19 adults, only 41.4% (= 4.4) of the females and 50% (= 4.5) of the males reach the beginning of sexual maturity. Supposing that each of those 4.4 females survives this period, which is rather improbable (only 1.8 females reaching the age of 200 days), the egg-production would be: for each female, 2.1 egg-pods (= 114 eggs), i.e. 487 eggs total. Considering that actually only 300 eggs are laid, 200 eggs are still supernumerary under the conditions of this breeding. However, under natural conditions, mortality would have been much heavier. Parasites such as *Idia lunata* or *Chortophila cilicrura* and fungous diseases would have decimated the egg-stage; lizards, birds, small mammals the larval stage; larger birds and mammals (including man) the adult stage (Fig. 1,1).

Table II.

Life-table of *Schistocerca gregaria* (bred in large outdoor cages, in the summer in Egypt).

| Age in days | lx | dx | qx | Factors decimating the stage in nature |
|---|---|---|---|---|
| 0 | 100.0 | 0.0 | 0.0 | Egg: Parasites: *Idia*, *Chortophila*, etc. |
| 5 | 99.0 | 1.0 | 1.0 | Fungous diseases |
| 15 | 94.0 | 5.0 | 5.1 | |
| 25 | 87.0 | 7.0 | 7.5 | |
| 35 | 61.5 | 25.5 | 29.3 | Larvae: Lizards and small mammals, |
| 45 | 34.5 | 27.0 | 43.9 | birds, man |
| 55 | 26.0 | 8.5 | 24.3 | |
| 65 | 21.5 | 4.5 | 17.3 | |
| 75 | 19.5 | 2.0 | 9.3 | |
| 85 | 19.5 | .. | .. | |
| 95 | 18.3 | 1.2 | 6.6 | |
| 105 | 16.0 | 2.3 | 12.6 | Birds and mammals |
| 115 | 13.8 | 2.2 | 13.7 | |
| 125 | 12.8 | 1.0 | 7.2 | |
| 135 | 12.2 | 0.6 | 4.7 | Man |
| 145 | 10.8 | 1.4 | 11.4 | |
| 155 | 9.1 | 1.7 | 15.7 | |
| 165 | 8.2 | 0.9 | 9.9 | |
| 175 | 7.3 | 0.9 | 11.0 | |
| 185 | 5.8 | 1.5 | 20.5 | |
| 195 | 3.5 | 2.3 | 39.6 | |
| 205 | 3.5 | .. | .. | |
| 215 | 1.6 | 1.9 | 54.3 | |
| 225 | 1.6 | .. | .. | |
| 235 | 1.6 | .. | .. | |
| 245 | 1.0 | 0.6 | 33.3 | |
| 255 | 0.6 | 0.4 | 40.0 | |
| 265 | 0.2 | 0.4 | 66.7 | |
| 275 | 0.0 | 0.2 | 100.0 | |

WIESNER & SHEARD[31] analysed the longevity of white rats of the stock of the Wistar Institute with variable diet under standard, but not optimal, conditions. As only the weaned animals are included in the life-table, the following corrections should be made: in the following computation, 5% are assumed to die as foetuses during the embryonic development; 32.66% died, according to the observations of the above authors, due to neglect on the part of the mother or to having been attacked by her during the first 30 days after birth. The embryonic period lasts 22 days (Fig. 1,II).

DEEVEY[32] objects to our conception of physiological tables. As we have already stated, they do not occur in nature. In addition, life expectation varies in every generation and in every environment, as we will see later. Also, mortality by endogenous senescence may set in at different ages in every population and this may raise or lower the trend of the survivor line. We agree, however, with DEEVEY that the mortality of the earliest stages is least known and that research should be concentrated on them. Our lifelong study of *Microtus guentheri* strongly supports this view.

We may conclude this paragraph with two life-tables concerning animals living under natural conditions[32].

**Table III.**
Life-Table of the Mountain Sheep *Ovis dalli dalli*, based on the known ages of 608 sheep before 1937

| $x$ age in years | $dx$ no. dying in interval out of 1000 born | $lx$ no. surviving out of 1000 born | $1000\ qx$ mortality rate per 1000 born at begin of age | $ex$ life expectation |
|---|---|---|---|---|
| 0.0–0.5 | 64 | 1000 | 54 | 7.1 |
| 0.5–1.0 | 145 | 946 | 153 | |
| 1–2 | 12 | 801 | 15 | 7.7 |
| 2–3 | 13 | 789 | 17 | 6.8 |
| 3–4 | 12 | 776 | 16 | 5.9 |
| 4–5 | 30 | 764 | 39 | 5.0 |
| 5–6 | 46 | 734 | 63 | 4.2 |
| 6–7 | 48 | 688 | 70 | 3.4 |
| 7–8 | 69 | 640 | 108 | 2.6 |
| 8–9 | 132 | 571 | 231 | 1.9 |
| 9–10 | 187 | 439 | 426 | 1.3 |
| 10–11 | 156 | 252 | 619 | 0.9 |
| 11–12 | 90 | 96 | 937 | 0.6 |
| 12–13 | 3 | 6 | 500 | 1.2 |
| 13–14 | 3 | 3 | 1000 | 0.7 |

In man, whose ecological mortality is greatly reduced by civilization, the comparison between the physiological and ecological life-tables is of lesser significance than in most other animals[34, 35]. Embryonic mortality is reduced to a very small number (artificial

## Table IV.
*Balanus balanoides* settling on cleaned rock in 1930 (age in months) [33]

| x | lx | 1000qx | ex |
|---|---|---|---|
| 0–2 months | 1000 | 90 | 12.1 |
| 2–4 months | 910 | 110 | 11.3 |
| 4–6 months | 810 | 62 | 10.5 |
| 6–8 months | 760 | 79 | 9.1 |
| 8–10 months | 700 | 114 | 7.8 |
| 10–12 months | 620 | 258 | 6.7 |
| 12–14 months | 460 | 174 | 6.7 |
| 14–16 months | 380 | 263 | 5.9 |
| 16–18 months | 280 | 179 | 5.7 |
| 18–20 months | 230 | 174 | 4.7 |
| 20–22 months | 190 | 526 | 2.4 |
| 22–24 months | 90 | 667 | 1.9 |
| 24–26 months | 30 | 667 | 1.8 |
| 26–28 months | 10 | 800 | 1.4 |
| 28–30 months | 2 | 1000 | 1.0 |

abortion excluded), as in all mammals, and the former high infant mortality is being gradually reduced. This is well expressed in Fig. 3.

### IDEAL LIFE TABLE OF MEN.

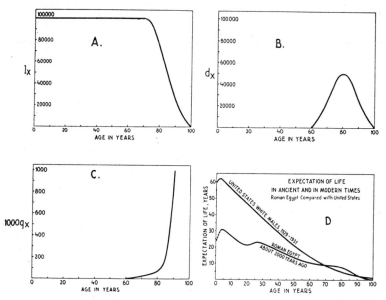

Fig. 3. Ideal (physiological) life-table curves of man and life-expectation in man. A–C as in Fig. 1. D. Life-expectation of man in ancient and recent times (D. after PEARSON).

In comparing now the theoretically possible types of life-curves of PEARL[6], we realize that all these types are only realizations of ecological life-tables, which have no connexion with inborn physiological life-curves. The real type of inborn physiological longevity is represented by the starving *Drosophila* flies in Fig. 2, in which the previous development stages are added as a line running closely parallel to the 100% living line until the adult stage is reached. All these possible types of PEARL are realized in ecological life-tables, types 4 and 3 being the most common, type 1 being a rather unusual exception induced by catastrophic changes in the environment.

We will now define some conceptions connected with the study of longevity:

(1) Life-span is the longest possible longevity of an individual within a species.
(2) Physiological longevity is the average longevity of individuals of a population of genetically homogeneous stock living under optimal conditions.
(3) Ecological longevity is the actual average longevity of the individuals of a population under given conditions.
(4) The life-table reflects the course of mortality in a given population. Its more important columns are: $l$ = survivorship at a certain age; $d$ = death cases occurring at each age-class; $q$ = death-rate at a given age-class, provided the entire population is composed of this age-class; $e$ = average life-expectance of an individual after having reached a given age.

The normal and usual conception of longevity is that called here ecological longevity.

## 4. Age-Distribution in Animal Populations

A greatly neglected study in animal ecology is that of age-structure of populations. Such studies are available almost exclusively for man. And in man the age-structure is intimately connected with the state of growth of a human population[36, 37]. Fig. 4 illustrates the growing populations as a pyramid, the youngest age-classes being always more numerous than the preceding age-class. As soon as the population becomes stable, the pyramid changes into a bell, where all lower and middle age-classes are more or less equal, the higher ages diminishing slowly. In a dwindling population the birth basis is reduced successively and the higher ages form a disproportionate part of the total population (urn-shaped). This latter case is the common picture of larger town populations (migration excluded). It will be of interest to study animal populations in this respect.

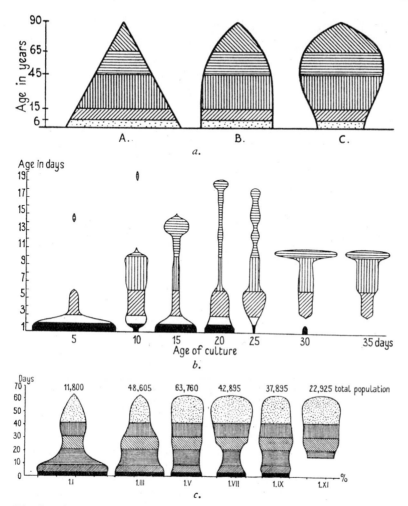

Fig. 4. Age-structure in changing populations: a. of man; b. of *Drosophila*; c. of honey-bee. A. Growing population; B. stable population; C. Contracting population.

The growth of animal populations within an empty space starts as unlimited growth until the environmental resistance begins to have an effect. And as we shall see, this influence of the environmental resistance begins at a rather early stage of population growth. A few types of population growth will now be analysed.

The most primitive type of population growth is that by simple division, met in most Protozoa and bacteria. It follows an exponential curve of the following formula:

$$P_n = P_1 \cdot 2^n,$$

where $P$ = population, $n$ = number of divisions (= generations), $P_1$ = population at the beginning of the experiment, $P_n$ = population after $n$ divisions. In this case no age-distribution occurs, because the individuality of the generation ceases at the moment of the division, and the products of division become automatically the members of the following "generation".

Fig 5, A illustrates this type of growth. None of the successive classes of the population live with those following, each class replacing the preceding one. However, no such curves of unlimited growth occur in nature. A real population growth of *Paramaecium multinucleata* has the following aspect[38]:

| Age of culture in days | Mean number of *Paramaecium* per c.c. | Mean number of divisions per day | Approximate % of conjugants induced |
|---|---|---|---|
| 0 | 1.4 | .. | .. |
| 1 | 5.1 | 1.85 | .. |
| 2 | 23.5 | 2.22 | .. |
| 3 | 108.8 | 2.21 | .. |
| 4 | 250.0 | 1.20 | 2 |
| 5 | 304.3 | 0.21 | 10 |
| 6 | 306.5 | 0.01 | 40 |
| 7 | 298.8 | −0.04 | 35 |
| 8 | 289.3 | −0.05 | 35 |
| 9 | 287.0 | −0.01 | 10 |
| 10 | 287.0 | 0.00 | 3 |
| 11 | 252.0 | −0.19 | 0.2 |
| 12 | 228.5 | −0.14 | .. |

The exponential growth is soon broken and changed into the general curve of organism growth, the logistic curve. This resistance is composed of two parts, working independently: destruction by enemies, diseases, accidents, etc., and inhibition of reproduction – in this case division. Highest population density creates an unfavourable environment and induces a new type of reproduction, i.e. conjugation. In our *Paramaecium* culture the latter factor alone worked. The following scheme may illustrate how destruction, as well as inhibition, induced up to a certain degree and increasing up to a certain point parallel to growing population density, will change the exponential into a logistic growth (Fig. 5, c).

The potential increase of any animal, the parents of which die when the young generations appear, is, in principle, similar to that of Protozoa. In many insects, tapeworms, fish, etc., we meet the same scheme, always inhibited in nature by intrinsic as well as extrinsic factors. In insect populations of overlapping generations

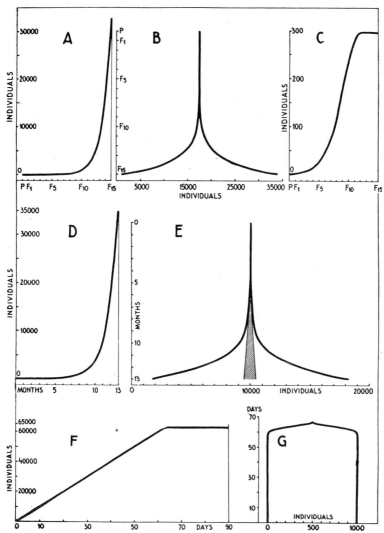

Fig. 5. Unlimited and limited growth and age-structure of animal populations. A to C, of Protozoans; D to E, of voles; F to G, of honey-bees.

A and D represent exponential increase; B and E age-stratification of exponential populations (shaded area in E: actual age stratification); C. actual population growth of *Paramaecium*; F and G: ideal population growth and age-stratification at saturation (= stability) in honey-bees.

the age-structure shows the same changes from pyramid over bell to urn in different growth-stages, as will be shown later in the chapter on *Drosophila*.

| Generation | Exponential growth by division | % destruction or inhibition per generation | Population after reduction as indicated in previous column |
|---|---|---|---|
| P | 1 | .. | 1 |
| $F_1$ | 2 | .. | 2 |
| $F_2$ | 4 | .. | 4 |
| $F_3$ | 8 | .. | 8 |
| $F_4$ | 16 | 5 | 15.2 |
| $F_5$ | 32 | 10 | 27.4 |
| $F_6$ | 64 | 15 | 46.6 |
| $F_7$ | 128 | 20 | 74.4 |
| $F_8$ | 256 | 25 | 111.6 |
| $F_9$ | 512 | 30 | 178.6 |
| $F_{10}$ | 1,024 | 35 | 232.2 |
| $F_{11}$ | 2,048 | 40 | 278.6 |
| $F_{12}$ | 4,096 | 45 | 296.5 |
| $F_{13}$ | 8,192 | 50 | 296.5 |
| $F_{14}$ | 16,384 | 50 | 296.5 |
| $F_{15}$ | 32,768 | 50 | 296.5 |

Mammals, birds, and other long-lived animals show a somewhat different type of population growth. In voles, for example, many generations of the offspring reproduce freely simultaneously with their parents. In 3 years' lifetime one couple of *Microtus guentheri* may give rise to 22 generations ($F_1$–$F_{22}$) and itself gives birth to 31 litters. The total number of offspring of every generation produced from the beginning of the first year will be, at the end of this year:

| Age in months | Number of individuals | Total population in the month |
|---|---|---|
| 12–13 | 2 | 2 |
| 11–12 | 6 | 8 |
| 10–11 | .. | 8 |
| 9–10 | 24 | 32 |
| 8–9 | 42 | 74 |
| 7–8 | 114 | 188 |
| 6–7 | 348 | 536 |
| 5–6 | 490 | 1,026 |
| 4–5 | 798 | 1,824 |
| 3–4 | 1,842 | 3,666 |
| 2–3 | 4,014 | 1,680 |
| 1–2 | 10,572 | 18,252 |
| 0–1 | 16,516 | 34,768 |

These figures are based on experimental breedings which yielded 6 young ones per litter, and births were observed at 35 days interval

($F_1$ after 25 days, first birth of neonate generation after 50 days). This potential growth is, of course, never realized. We will have occasion to compare it with actual population dynamics of that species in a later chapter.

The curves expressing age-distribution and potential growth require some comments. The age-distribution shows an absurdly large and ever rapidly extending basis. No such age-distribution is ever met with in nature. Even within a rapidly growing population, the part shaded in Fig. 5 E would be realized at the utmost. This means that heavy infant mortality must occur in nature. It is verified for voles, where the young ones are destroyed by enemies and unfavourable factors in a percentage much surpassing their numerical share[39, 40].

The same type holds for all mammals and birds, etc. with the only difference that the factor of increase is much lower and the rate of increase much slower. The elephant may serve as a theoretical example with 100 years longevity and 4 young ones on the average $(m. = f.)$:

| Generation | P | $F_1$ | $F_2$ | $F_3$ | $F_4$ | $F_5$ | $F_6$ | $F_7$ | $F_8$ | $F_9$ | $F_{10}$ |
|---|---|---|---|---|---|---|---|---|---|---|---|
| Rise of population | | 2 | 4 | 8 | 16 | 32 | 64 | 128 | 256 | 512 | 1,024 | 2,048 |

In the elephant, 10 generations extend over more than 500 years, whereas in voles over at least 16 months, in Protozoa over a day only.

A particular type of restriction may still be discussed. In the beehive the number of reproducing individuals is reduced to one queen only. Similar cases occur in wasp, ant, and termite populations. The daily oviposition rate of such a queen depends highly upon environmental factors such as weather, amount of nectar available, status of population within the hive, etc. The author has shown that in bees, where an annual seasonal trend in the hive population can be observed, the age-structure of the changing population has the same shape, from pyramid over bell to urn, as that of human populations[19, 41].

The growth of honey bees in primitive plaited tubes, which must be very similar to those in wild bee nests, differs considerably. We followed their development for one year in Ankara, Turkey[42]:

Oviposition started on the 6th March and continued to 19th October. During these 228 days oviposition was interrupted three times (three days in March, two days in June, four days in September). The main oviposition period was from early April to early June. The very marked spring peak was about the middle of April. The total egg-production was 34,000 (as compared with 220,000 at Kiryat Anavim, near Jerusalem, in a modern hive)[19]. We must, of course, take into consideration the four and a half months of hibernation on the Anatolian plateau, which are lacking in the sunny South. At Ankara the strict dependency of the development of the hives on environmental conditions is conspicuous. We have here a very definite interplay between these conditions and their utilization by the bees in various degrees. We regret that this is not the place to discuss this interplay in detail. Particulars of the development are given in Table V and in Figure 6.

Table V.
Population Counts of Ankara Native Hives (BODENHEIMER, 1942)

| Date | Living adults | Dead adults | Eggs | Larvae | Pupae | Total | Honey cells | Pollen cells | Brood cells | Empty cells | Total cells |
|---|---|---|---|---|---|---|---|---|---|---|---|
| 3.I.39 | 907 | 79 | – | – | – | 907 | 4702 | – | – | 14347 | 19049 |
| 23.I.39 | 2192 | 97 | – | – | – | 2192 | 5402 | – | – | 10766 | 16168 |
| 10.II.39 | 1967 | 210 | – | – | – | 1967 | 1496 | – | – | 12699 | 14105 |
| 3.III.39 | 2245 | 179 | – | – | – | 2245 | 6890 | – | – | 14907 | 21797 |
| 25.III.39 | 3372 | 113 | 110 | 10 | 64 | 3556 | 7560 | 158 | 184 | 25680 | 33672 |
| 18.IV.39 | 3502 | 740 | 3130 | 2074 | 1212 | 9918 | 2699 | 483 | 6416 | 10357 | 19955 |
| 10.V.39 | 4057 | 302 | 1041 | 1945 | 3464 | 10507 | 3700 | 1031 | 6450 | 17967 | 29148 |
| 2.VI.39 | 7986 | – | 964 | 1148 | 3509 | 13607 | 3902 | 644 | 5721 | 15525 | 25792 |
| 11.VII.39 | 10427 | – | 685 | 64 | 3267 | 14443 | 3699 | 428 | 4016 | 14287 | 22430 |
| 21.VII.39 | 6996 | – | 761 | 720 | 1668 | 10145 | 1808 | 135 | 3149 | 19493 | 24585 |
| 31.VII.39 | 3263 | – | 179 | 612 | 1482 | 5536 | 12008 | 326 | 2263 | 13754 | 28451 |
| 20.VIII.39 | 2619 | – | 1381 | 451 | 443 | 4984 | 8755 | 233 | 2275 | 29380 | 40644 |
| 9.IX.39 | 2410 | – | 1565 | 455 | 1386 | 5816 | 9697 | 253 | 3406 | 15059 | 28415 |
| 29.IX.39 | 1988 | – | – | 12 | 443 | 2443 | 8062 | 206 | 455 | 18108 | 26838 |
| 19.X.39 | 2863 | – | 35 | 45 | 291 | 3234 | 11052 | 106 | 371 | 13187 | 24716 |
| 8.XI.39 | 4305 | 35 | – | – | – | 4305 | 13001 | 341 | – | 4738 | 18080 |
| 28.XI.39 | 3610 | 12 | – | – | – | 3160 | 12259 | 41 | – | 20045 | 32345 |
| 18.XII.39 | 4626 | 11 | – | – | – | 4626 | 17352 | 254 | – | 10748 | 28354 |
| 18.I.40 | 5576 | 13 | – | – | – | 5576 | 9509 | 74 | – | 21803 | 31386 |
| 18.II.40 | 6254 | 10 | 424 | 125 | 346 | 7149 | 9107 | 595 | 895 | 15687 | 26224 |
| 18.III.40 | 5743 | 4 | 119 | 81 | 80 | 6023 | 4862 | 339 | 280 | 21846 | 27327 |
| 12.IV.40 | 4471 | – | 318 | 328 | 341 | 5458 | 4406 | 193 | 987 | 10732 | 25318 |
| 3.V.40 | 1781 | – | 187 | 247 | 611 | 2826 | 668 | 146 | 1045 | 15100 | 16959 |
| average | | | | | | | 7363 | 265 | 1675 | 15888 | 25191 |

Assuming that in a modern hive, under constant suboptimal conditions, which last for some time, the queen lays 1,000 eggs per day, the growth curve will differ greatly from the exponential one. The reason is, of course, the restriction in the number of reproducing individuals. The mortality in bee-hives at almost optimal conditions (even temperature, brood care, etc.) seems to be negligible. Through the 21 days of development, the 10 days of nurse-bees, and the 10 days of house-bees, the population would show a daily increase of 1,000 eggs. The last 21 (—24) days of the field-bee would show on the 62nd to 66th day the beginning and end of mortality. Fig. 5F and G shows this type of increase and age-distribution. The formula for this type of growth would be as follows:

During the first 63 days of growth: $P_n = a.e$

After that period: $P_n = 63.e,$

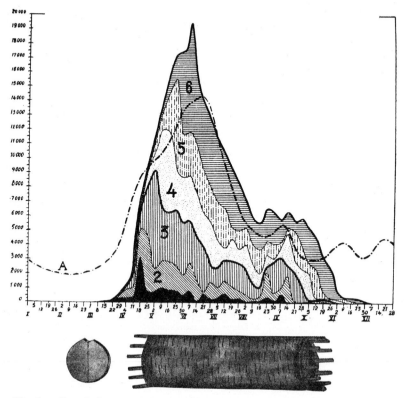

Fig. 6. Population trend in the native plaited bee-hive (below) at Ankara. Calculated population: Black. eggs; 2. larvae; 3. sealed brood; 4. nurse bees; 5. house bees; 6. field bees. Dotted line (A): Actually counted population showing the numbers of actually hibernating adults. Observe the delayed summer peak, showing the slow beginning of the annual increase (From BODENHEIMER, 1942).

where $a$ = age of population in days, $e$ = constant daily oviposition rate.

The age-distribution after the 63rd day is of the typical bell shape of a stagnant population.

The age-structure of a population serves, if the bionomic constants for the species are known, as an important indicator of the character of the life-table of the population concerned.

### 5. Relation between the Three Ecological Ages

From a physiological point of view the relative age of an organism may be divided into many categories, depending on growth of weight, size, linear growth, etc. The ecologist is interested mainly in age in connexion with reproduction. The life-span will thus be divided into development, lasting from fertilization of the egg until first birth; into the reproductive period, lasting throughout reproduction; into the post-reproductive period of senescence. Only for very few animals do we possess material based on sufficiently large-scale observations to enable us to get significant averages. A few examples of relative age are quoted below:

**Table VI.**
Relative duration of ecological ages in some animals.

|  | Development | Reproduction | Post-reproduction | Average longevity in days |
|---|---|---|---|---|
|  | % | % | % | % |
| *Homo sapiens,* |  |  |  |  |
| ,, ,, average | 22.1 | 51.7 | 26.– | 17,430 |
| ,, ,, extreme | 12.9 | 45.2 | 41.9 | 19,254 |
| *Rattus norvegicus,* |  |  |  |  |
| ,, ,, average | 24.8 | 20.6 | 54.6 | 1,104 |
| ,, ,, extreme | 11.7 | 37.6 | 50.7 | 1,374 |
| *Drosophila melanogaster* | 41.6 | 54.2 | 4.2 |  |
| *Pieris brassicae* | 95.9 | 4.1 | .. |  |
| *Schistocerca gregaria* | 50.5 | 17.5 | 32.0 |  |
| *Ephemeridae* | 99.86 | 0.14 | .. |  |
| *Panolis flammea* | 98.9 | 1.1 | 2.8 |  |
| *Periplaneta americana* | 69.9 | 27.2 | .. |  |
| *Tenebrionides mauretanicus* | 29.9 | 46.5 | 24.3 |  |
| *Trogoderma granarium* | 83.0 | 11.9 | 5.1 |  |

The main difference between mammals and insects in this group is the increased share of development and the reduced part of reproductive and post-reproductive period in the insects. This is conditioned mainly by the shift of the total feeding period into the larval stage (Ephemeridae), whereas in species with a long-lived adult stage (Tenebrionidae) the share of each period resembles closely that of both mammals. The difference is therefore not of prime importance.

The use of this type of longevity is, of course, restricted when we observe the great number of animals (many mammals and birds, and some insects such as Tenebrionidae or Dytiscidae), which have a more or less extended annual reproductive season (this is true of multi-annual species only) with an annual reduction of the gonads to an infantile condition during the infertile remainder of the year. In ants we have very special conditions during the early colony-foundation of the future queens, i.e. the just fertilized females. In some primitive Ponerinae ants in Australia the future queens in clausure break out for predation whenever they are hungry, returning then to their claustrum to continue colony foundation. In all higher ants, in contrast, the claustrum is never broken. The future queen starves and oviposits and takes care of her brood. She starves, eats her own fat-body, flight-musculature, part of the eggs and larvae which she has produced, feeds the larvae with another fraction of the eggs and larvae produced, until finally the first nanic workers appear, feed their queen and themselves and the remaining brood, thus actively supporting the beginnings of the colony foundation.

The situation becomes still more complicated in many molluscs, where the sexes alternate in the same individual according to season, situation, etc.

Longevity calls for far more biological research before we shall be able to bring the facts into really comparable schemata.

## II
## THE LIFE-HISTORY AND ITS ECOLOGICAL INTERPRETATION TO-DAY

One of the American pioneers in ecology once said that the life-history of the individual species will always remain the basis of animal ecology. This statement is no doubt true. The question, however, as to what should be understood by life-history has recently undergone many changes. It is not any longer restricted to the description of stages and of parasites and predators. The abundance of recent methodological achievements is large. Only a few of the more important implications will here be discussed.

Exact determination, description of all stages and ages, observations on their duration in the field as well as under experimental conditions in the laboratory, observations on reproduction and fertility on normal and occasional food (qualitatively and quantitatively, and especially on nutritional mortality), and statistical measurements of environmental resistance are still standard requirements and there is no need to dwell on them. We propose here to deal mainly with some problems of physical ecology.

It is high time for ecology to follow the lead of J. DAVIDSON[1] and abandon the indiscriminate use of the term "climate" reserving it for what it actually is: – an abstraction, an average of weather changes in any place or region, with no relation whatsoever to the actual weather of any special season. In this sense, climate is an important concept in zoogeography, but has very little to do with the daily and hourly changes of the weather in any environment, which influence the annual or seasonal abundance of many animals to so great an extent.

When we go into detailed studies, it becomes evident that the relations between abundance and weather are extremely manifold and complex. We mention here one of the most detailed weather studies, that of KINCER[2] on the Cotton Boll Weevil *(Anthonomus grandis)* in Texas. The following weather phases were regarded as most important for the damage done by this pest in the concurrent season, namely: the relative humidity in June/July of the current season; the lowest winter temperature of the preceding winter; the relative humidity of July/August of the preceding year; and the percentage of cloudiness in June/September of the preceding year. Especially important is the relative humidity of both seasons. The weevils deposit their eggs in the young bolls, and the hatching larvae feed on the interior of the bolls. When punctured, many young bolls drop to the ground and the larvae may die from intense heat or from the drying up of these bolls. On the other hand, moist, cloudy, rainy weather favours a rapid increase in numbers from the several generations of the growing season. In Texas, the multiple correlation coefficient for the combined factors, mentioned above, and for weevil damage is R = 0.934, an exceptionally high positive correlation.

## 1. The Physical Ecology of the Life-Cycle

The duration of the life cycle depends on many environmental factors, of which the chief one is usually temperature. Humidity, quantity, and quality of accessible food, etc., are other important factors, but for the normal life-cycle of most animals their importance is small as compared with that of temperature. Other factors, however, may need to be used for local corrections of the pure temperature dependency.

We do not intend here to take up again the discussion of the curve approaching nearest the experimental data of temperature, humidity, etc., dependency of development[3, 4]. For the intermediate zone, hyperbola and parabola, as well as catenary curve, are all in rather close agreement with the data. The inhibition of development above the point of shortest development must be found experimentally for the last two types of curves, whereas it may easily be entered as a correction into the hyperbola, wherever this upper zone is needed. The three distinct advantages of the equilateral hyperbola are:

(1) The extremely simple way of computation, an important fact considering the mathematicophobia of most biologists.
(2) The easy way of getting the basic data for calculation and of improving the curve by additional data.
(3) Last, but not least, only the hyperbola furnishes us with the value for the lower development threshold, which value, even if only approximate, has proven to be of highest importance in ecological conceptions and calculations. This value is not obtained by any other formula.

The corrected temperature-sum rule is, within a certain vital range, certainly a very close description of the facts observed. Its exactitude and value should, however, not be exaggerated, as every mathematical approach in biology serves only as a short description of biological observations. Neither should the hyperbola be interpreted as the best physiological solution to the problem of temperature dependency. For ecological purposes, however, it has proven invaluable, not only because its immediate values are of greatest importance and interest, but also because many other applications of this simple formula may be advantageously made.

Some of these applications may briefly be mentioned:
(1) The thermal constant is in certain cases the limiting factor in northern distribution. The area of *Gueriniella serratulae* is limited

wherever the local annual sum of efficient temperature falls below the thermal constant (c = 13.3° C thermal constant = 1188 day-degrees)[5].

(2) In certain species splitting into variations, or subspecies, has occurred in the zone in which the thermal sum divides the area of the species into areas with one or two generations per year or one generation in two years, as is the case within the "Rassenkreis" of the pine shoot moth *Evetria buoliana*[6].

(3) The calculation of the theoretical local life-history by means of a hyperbola has proven to be of greatest value and economy when correct, or, if incorrect, has stimulated research on other limiting factors. The moth *Sitotroga cerealella* e.g. showed at Degania in the Upper Jordan Valley a considerably longer development than could be expected from the hyperbola based upon breedings on the coastal plain. Investigation revealed that the water content of the wheat was 8% in Degania, as against about 12% on the coastal plain, which explained the difference adequately[7].

(4) The annual fluctuations of the sum of local effective temperatures above the thermal constant have been of great use in explaining the local fluctuations of important pests.

(5) The hyperbola permits application, not only for temperature-dependency of development, but also for activity, oviposition, etc.[8, 9, 10].

The first primitive form of the hyperbola as symbolic of the local life-cycle has undergone many improvements. Its actual shape will be here exposed.

One of the best studies in the search for a suitable expression of the temperature correlation to length of development was made by WADLEY[11] concerning *Toxoptera graminum*. He tried a great number of possible mathematical curves to find the one best suited to the experimental data. The well-known curves of ARRHENIUS and of VAN'T HOFF ($Q_{10}$) were wholly unsatisfactory. A double logarithmic curve and that of a cubic parabola did not fit quite as well as the straight line (i.e. the inverse of the hyperbola). WADLEY concluded, however, that a very flattened S– (logistic) curve, almost straight in its long middle part, may fit better. Yet all curves with a more or less satisfactory empirical fit are rather complicated to calculate (true also of PRADHAN's[12] formula), and are beyond the reach of the average ecologist. None of them, apart from the hyperbola, give the ecologically important threshold of development, while the upper turning point in every case has to be ascertained experimentally. The easy handling and calculation of the equilateral hyperbola makes it a very useful instrument in the hands of the ecologist, at least for a first approach. These conclusions coincide completely with our own.

It should be added that in heliothermic animals, where the body-temperature is raised greatly during the day through sun radiation, beyond that of the average environmental temperature, the hyperbola cannot possibly be applied for their life in nature.

Whereas most of our earlier calculations for insects were based on the development only, ended by the appearance of the adult insect, we now extend the development stage until sexual maturity.

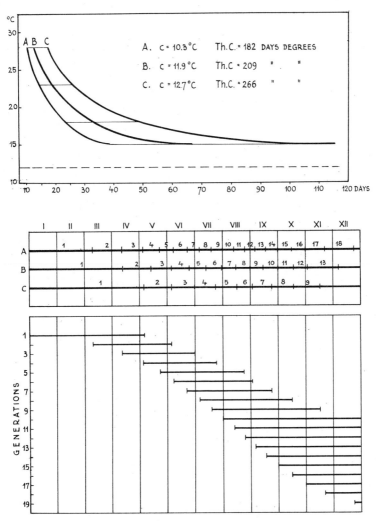

Fig. 7. Triple band of hyperbolae (I); local life-histories (II) of the three hyperbolae, and succession of generations (III) at Tel Aviv calculated and observed for the Red Spider *Anychus latus*.

And whereas in insects the fertilization of the eggs precedes oviposition, as a rule, only for a very short period, we may define the period of development as lasting from birth (oviposition of parents) to first oviposition. This does not interfere with the possibility of constructing, if desirable, separate hyperbolae for each of the development stages (e.g. egg, each larval and the pupal stage, preoviposition period). This hyperbola comprehends therefore the whole prereproductive period of the animal. But the animal does not die at the end of reproduction; a second hyperbola may be constructed for the average longevity of the individuals of any generation from birth to death. The construction of an intermediate hyperbola seems generally essential. For epidemiological purposes the average life-cycle of any generation is represented by a hyperbola which runs from birth to the end of the first half of normal oviposition (with regard to quantity of eggs, not to oviposition period). For other purposes it may be preferable to calculate the intermediate hyperbolae for the end of the reproductive period. In insects, however, the post-reproductive period of senescence is often negligibly short.

This triple band of hyperbolae permits the reconstruction of the local life-history with its overlapping of generations and its mixture of hibernating or estivating generations, to a hitherto unknown degree of exactness and detail.

The representation of the life-history of the Oriental red spider *Anychus latus* may serve as an example (Fig. 7). The agreement of this calculated life-history with the actual one, which has been observed for some years by my associate KLEIN[13], is almost perfect. This agreement is the more significant as the relatively large number of annual generations would increase any important deviation enormously. The local life-history, as well as the local sequence of generations, have been worked out satisfactorily for quite a considerable number of citrus and other insects in Israel[14, 15, 16].

## 2. Life-Cycles of Some Lady Beetles

Further improvements may be added for more complicated life-histories, as will be shown now for a family with especially complicated life-cycles, the Coccinellidae.

The simple life-cycle is often not realized, and it is desirable to introduce corrections within the simple distributional life-chart.

We have had an opportunity to study the life-cycle of the common seven-spotted lady beetle *(Coccinella septempunctata)* in four different zoogeographical regions[17]. Its threshold of development is at 12.6° C, its thermal sum for hyperbola B is 379 days-degrees (A: 97, C: 459). The favourable zone is between 16 and 20°C and 65 to 80%

R.H. Temperatures above 23° C have a detrimental effect on all stages and their activity. Oviposition, feeding, etc. are interrupted in the adults; no development is continued in the larvae. The beetles begin to be inactive at temperatures below 14° C. The number of calculated generations including the times of interruptions (in brackets) are given in the following table:

**Table VII.**
Period of development (interruption in brackets) and number of generations of *Coccinella septempunctata* L.

| Regions and Localities | Month favourable to development | Generations by Hyperbola | | |
|---|---|---|---|---|
| | | A | B | C |
| A. Euro-Siberian | | | | |
| Paris | V-IX (X-IV) | 1.7 | 1.5 | 1.2 |
| Pinsk | V-IX (X-IV) | 1.8 | 1.7 | 1.3 |
| B. Mediterranean | | | | |
| Genova | IV-VI (VII-VIII), IX (XII-III) | 1.4 +1.3 | 1.2 +1.0 | 1.0 +0.9 |
| Izmir | IV-V (VI-VIII), IX-XI (XII-III) | 1.0 +1.7 | 0.8 +1.4 | 0.7 +1.1 |
| Tel-Aviv | II-V (VI-IX), X-XII (I) | 1.8 +1.8 | 1.5 +1.5 | 1.2 +1.2 |
| Jerusalem | III-VI (VII-VIII), IX-XI (XII-II) | 2.0 +2.0 | 1.6 +1.6 | 1.4 +1.4 |
| C. Irano-Turanian | | | | |
| Ankara | V-VI (VII-VIII), IX-X (XI-IV) | 1.1 +0.7 | 0.8 +0.6 | 0.7 +0.5 |
| D. Saharo-Sindian | | | | |
| Cairo | III-V (VI-IX), (X-XI) (XII-II) | 1.9 +1.4 | 1.6 +1.1 | 1.3 +0.9 |
| El-Arish | III-V (VI-IX), X-XII (I-II) | 1.4 +0.7 | 1.2 +0.5 | 1.0 +0.4 |
| Dakhla Oasis | XI-III (IV-X) | 1.6 | 1.4 | 1.1 |

The life-histories in the various zoogeographical regions may be described as follows:

(A) Euro-Siberian region. – During five consecutive months (May to September) average temperatures are above the development zero. Winter temperatures fall considerably below the threshold of development. During no month does the average temperature rise above 23° C. No semi-estivation is therefore enforced. For three consecutive months (June to August) climatic conditions are within the favourable zone. The sum of effective temperatures permits the full development of one annual generation, but scarcely, in some districts, the partial development of a second

Average monthly temperatures of some localities
within the area of distribution of *Coccinella septempunctata* L.

| Regions and Localities | Month | | | | | | | | | | | |
|---|---|---|---|---|---|---|---|---|---|---|---|---|
| | I | II | III | IV | V | VI | VII | VIII | IX | X | XI | XII |
| A. Euro-Siberian | | | | | | | | | | | | |
| Paris | 2.3 | 3.6 | 5.9 | 9.9 | 13.0 | 16.5 | 18.3 | 17.7 | 14.7 | 10.1 | 5.8 | 2.7 |
| Pinsk | −5.4 | −4.5 | −0.4 | 6.9 | 13.8 | 17.6 | 19.0 | 7.7 | 13.1 | 7.0 | 0.9 | −3.8 |
| B. Mediterranean | | | | | | | | | | | | |
| Genova | 7.5 | 8.7 | 11.0 | 14.0 | 17.3 | 21.1 | 24.1 | 24.1 | 21.3 | 16.7 | 11.8 | 8.6 |
| Izmir | 8.4 | 8.3 | 11.3 | 15.3 | 20.0 | 24.5 | 27.4 | 27.3 | 22.6 | 18.8 | 13.9 | 10.1 |
| Tel-Aviv | 12.8 | 13.5 | 15.5 | 17.9 | 21.5 | 24.5 | 26.4 | 26.7 | 25.5 | 22.6 | 18.7 | 14.7 |
| Jerusalem | 8.3 | 8.9 | 12.7 | 15.4 | 20.0 | 22.0 | 23.2 | 23.2 | 22.0 | 20.2 | 15.8 | 10.4 |
| C. Irano-Turanian | | | | | | | | | | | | |
| Ankara | −0.4 | 0.8 | 5.5 | 11.3 | 16.4 | 19.6 | 23.4 | 23.5 | 18.5 | 13.7 | 7.5 | 2.4 |
| D. Saharo-Sindian | | | | | | | | | | | | |
| Cairo | 10.9 | 12.3 | 15.2 | 19.0 | 22.6 | 23.4 | 26.6 | 26.3 | 23.9 | 21.6 | 17.3 | 12.7 |
| El-Arish | 10.6 | 11.3 | 13.5 | 17.3 | 20.7 | 24.0 | 25.4 | 25.2 | 23.4 | 20.9 | 16.8 | 12.3 |
| Dakhla-Oasis | 13.0 | 14.2 | 18.3 | 24.3 | 28.0 | 30.7 | 30.8 | 30.7 | 28.7 | 25.0 | 19.4 | 14.5 |

one. Food conditions are unfavourable for the development of a second generation.

The typical life-history of *Coccinella septempunctata* L. within the Euro-Siberian region is as follows: One generation develops during summer, coinciding with the season of aphid abundance. The young beetles hide for about seven months between leaves or in crevices. Climatic as well as food conditions must be considered as favourable.

(B) Mediterranean region. – For five to zero months the average temperature falls below the development threshold, but never considerably. Average temperatures surpass 23° C. for 2 to 4 months during summer, thus enforcing semi-estivation. In the two intervals, during spring and autumn, no two consecutive months fall into the favourable climatic zone. The sum of effective temperatures always permits the development of one full generation each in spring (April to May-June) and again in autumn (usually September–November). In a few cases an abortive partial second generation may develop each in spring and in autumn, as reported for Israel, but these partial second generations are subject to very heavy mortality and are of no importance for the maintenance of the species.

On the whole, climatic conditions are not very favourable for the development of *Coccinella septempunctata* L. Both development periods coincide with the presence of sufficient food, especially that of the spring generation. Migrations to high mountains have been amply reported, likewise disappearance or at least inactivity, extreme scarcity and absence of development stages during the summer interval (temperature above 23° C). It is quite probable that part of the summer-migrants to the hills remain there until the following spring, whereas others, probably most, are reactivated in autumn and are not subjected to a true hibernation, but only to sluggishness due to low temperatures. On sunny days, they are active also during winter. If all or only most beetles surviving until autumn have died, the summer migration remains to be studied.

The hibernated beetles start oviposition in March. Their eggs

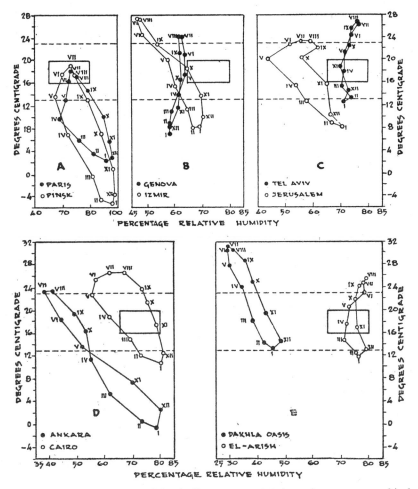

Fig. 8. Ecoclimograms of *Coccinella septempunctata* in various zoogeographical regions of the Palaearktis (From BODENHEIMER). The lower dotted line indicates the lower temperature threshold, the upper one the begin of uncomfortable high temperatures. The rectangle indicates the zone of optimal climatic conditions.

form the first generation, the adults of which appear in April. A small number of the beetles of the first generation oviposit in May. Less than 1% of these eggs develop successfully into adults, and most of these beetles die before the beginning of oviposition. This IIA generation is of no importance in the maintenance of the species. The great majority of the first generation, which matured in April, estivates. Oviposition and activity is resumed in the autumn (late September), and the beetles die soon afterwards. The second generation develops from their eggs in October. Some of these adults oviposit, forming a IA generation. A small proportion only of the eggs of this IA generation develops successfully into beetles which lay very few eggs during the winter. These eggs do not develop. The few survivors participate in the formation of the normal first generation. The generation IA is therefore also of little importance in the maintenance of the species.

The brothers of those parents which did not oviposit in October hibernate. They are active, with interruptions, during the winter, especially on sunny days, and even oviposit occasionally. From the eggs of the hiberating second generation, the first generation develops in March. We thus observe two annual generations in the coastal plain in spring and autumn which develop more or less sterile sidelines, one each in summer and winter.

(C) Irano-Turanian region. – For six months temperatures are considerably below the threshold of development, for two summer-months above 23° C. Climatic conditions are generally unfavourable to *Coccinella septempunctata* L. at Ankara. The monthly combinations of average temperature and air humidity are all far beyond the favourable zone of development (16–20° C, 65–80% R.H.). Furthermore, not one month during the years analysed (1926–1939) falls within this favourable zone. Climatic conditions are still much more unfavourable in autumn than in spring. Exposed to the general climate of Ankara, the survival would probably be small indeed. This is in agreement with observations. The more favourable microclimate of the estivo-hibernacula, especially their high humidity and their protection against wind and sun-radiation, is essential for the mass survival of the ladybird. The sum of effective temperatures in late spring (May-June) permits the development, but not the full maturation, of one generation. This spring period coincides with the season of aphid abundance. The sum of effective temperatures during the autumn period (September–October) is quite insufficient for the formation of one full generation.

The life-cycle comprises one spring generation which migrates to mountain-tops in late summer and remains there beneath stones in aggregations in semi-torpor. Others remain on the top of the mountains in cushions of plants such as *Astragalus*, but almost all die, as do those which remain below on the high plateau. The reasons why the beetles do not leave their estivo-hibernacula in autumn are unknown, but probably not directly dependent on conditions of temperature or humidity.

(D) Saharo-Sindian region. – It is obvious that the species cannot exist permanently within the real desert, where rains may be absent for one or more years and where low air humidities are very unfavourable even during winter and spring, when temperatures are favourable to the development of two to three generations. But the desert fringe (El Arish) still offers favourable conditions for development with a thermal sum permitting the development of one or two generations during the cool period, provided that rains and, consequently, vegetation with aphids are available. Even under the still apparently not extremely unfavourable conditions of Cairo, with relatively high air humidity, the species is unable to build up populations of the size known from the other regions, discussed before, and is extremely rare.

Within the whole range of climates which we have analysed there are few tropical climates at higher altitudes which would permit more than two to three annual generations for *C. septempunctata*.

Similar seasonal limitations are winter interruptions only. There are other ways of passing unfavourable seasons. *Chilocorus bipustulatus* and *Cryptolaemus montrouzieri* undergo in this country a very heavy summer mortality, which reaches almost 100% from June to August. Among the 4–5 generations, only a few adult beetles of 2 generations pass the summer, whereas the adults of 3 generations passing the winter have a much lower mortality.

Turning now to the local life-histories of some Coccinellid-beetles in Israel, we find the most archaic type realized in *Lithophilus marginatus*[14]. This predator of *Icerya purchasi* has a prolonged adult diapause. The beetles hatch from May to September and remain sexually immature until the following spring. The total adult longevity was 523 days on the average, including 170 days oviposition, with 230 eggs per female. The maintenance of the species is thus well safeguarded. The poor season (autumn and winter) is passed by the immature adults, and adult longevity is long enough to permit oviposition throughout spring and summer. The development of a fraction of the offspring is achieved even after a sequence of a few unfavourable months.

In *Epilachna chrysomelina*[15], a pest of cucurbitaceous plants, there are six months of development and six months of hibernation. Whereas oviposition and other activities of the adult beetles stop at about 20° C, the threshold of development falls at 11.3° C. No development was observed from after the end of October to early May. From early May to the end of October, five generations were observed in breedings, 3–5 probably develop in nature, adults of the third to fifth generation hibernating. The heliothermy of this species in the larval, as well as in the adult stage, tends to increase the utilization of the thermal possibilities. Two generations have been observed in South Africa by GUN. This agrees well with our formula:

|  | Annual generations | | |
|---|---|---|---|
|  | Hyperbola A | Hyperbola B | Hyperbola C |
| Capetown | 1.3 | 0.7 | 0.5 |
| Wellington | 2.0 | 1.1 | 0.8 |
| Simonstown | 2.7 | 1.5 | 1.0 |

In the subfamily Coccinellinae, *Hyperaspis polita*[14] and the related *Oxynchus marmotta*, both predators of mealy bugs, approach the *Lithophilus* type closely. Oviposition is observed from May to August. The old beetles die in July-August, whereas the beetles of the current year hibernate. They do not seem to start oviposition in the same year until next May.

A female lays about 400 eggs. From each 100 eggs there grow into adults in

| V | VI | VII | VIII | IX | X |
|---|----|-----|------|----|---|
| 23 | 55 | 18 | 25 | 11 | 4 |

individuals.

In *Scymnus quadripunctatus* a continuous series of five annual generations is observed[14]. Beetles of the fourth, but mainly of the fifth generation, hibernate. This last generation develops in 217 days. Whereas winter is passed mainly in the adult stage, imaginal longevity, from spring to autumn, of the first to fourth generations, is rather short, and rarely surpasses four to six weeks.

The armoured-scale ladybird *Chilocorus bipustulatus*, has 4–5 successive annual generations[18]. The 1–2 summer-generations which follow the spring maximum have a much reduced egg-number and undergo a rather excessive mortality. The adults of 2–3 generations may survive the winter, but it is mainly those of the last pre-hibernal generation which do so. This behaviour is an approach to the complete summer diapause in *Coccinella*, but adult longevity is not yet extended sufficiently to safeguard the persistence of the species throughout the summer.

In *Cryptolaemus montrouzieri*[19], which was introduced into Israel for the purpose of controlling *Pseudococcus citri*, no acclimatization was achieved. The same is true for the species in Egypt, where acclimatization was attempted on a much larger scale (cf.[20]) Four to six annual generations undergo heavy climatic decimation in winter and especially in summer. Insufficient host density and lack of activity, even below the threshold of starvation at the elevated summer temperatures, bring about this result.

*Novius cardinalis*[21], the classical enemy of *Icerya purchasi*, is also, in its dynamics, the best adapted of all ladybirds studied. Five to nine generations follow each other without interruption, development extending to 3–4 months only, even in winter. Adults of 2–4 generations may pass the winter. Mortality is often high after heavy desert winds in the spring, whereas the summer heat is not very noxious. Host-density is a major factor determining the *Novius* populations.

The coincidence between all these life-histories and seasonal numerical fluctuation at Tel Aviv, physical factors (temperature, precipitations, humidity) and seasonal numerical host-fluctuation has been studied in our "Citrus Entomology"[14].

We may conclude that this method of describing life-histories, if properly executed, shows a close correlation with the actual life-history. It illustrates all points of major importance for physical ecology, at least with regard to temperature. It illustrates well the often very complicated overlapping of generations in winter or at any other season. It has shown an easy and competent agreement with life-cycles calculated for other regions and localities. The method is easily executed and fulfils many desiderata unrealized hitherto.

It would, however, be premature to expect that such mathematical extensions of a life-history in one locality could eo ipso be applied for its life-cycle in other regions. Our regional life-histories of *C. septempunctata* offer sufficient proof that many factors cooperate and co-act to bring about each result. This cooperation may be similar in many localities, but it may differ in others. Yet every sizable deviation from such a mathematical extension will automatically draw attention to other factors or to factor limits so far overlooked, or to differences in the physiological and/or genetical character of various populations of the same species, etc., which would otherwise have escaped notice for a long time. The application of these and other related methods can therefore be recommended, so long as they do not claim that the results obtained are invariably correct. Our repeated application[14, 15, 16,] of such methods has been fertile, even if only in laying bare important deviations. Like all knowledge, collected bits of information form a growing stream which may change its course eventually because of some additional solid research.

### 3. Applications of the Climogram (Paired Factors)

In animals which really have many annual generations – as is the case with *Anychus latus* – it will be possible to evaluate the optimal climatic conditions for the species by simultaneous observation of seasonal abundance. One must take care, however, to look less at the actual figures than at the quotient of increase. In a sequence of generations:

| Generation | 1 | 2 | 3 | 4 | 5 | 6 |
|---|---|---|---|---|---|---|
| No. of individuals | 0.1 | 2 | 5 | 20 | 120 | 130 |
| Quotient of increase | | 20.0 | 2.5 | 4.0 | 6.0 | 1.1 |

the transition from the first to the second generation brings about the decisive increase. The absolute population abundance veils this fact very often. Anyhow, only the weather at the decisive periods of increase and not at the population peak should be regarded as optimal. Where only one to two annual generations occur, observations should be made for a series of years. But instead of the epidemiological approach we may choose also a zoogeographical analysis. In that case climograms of high, medium and low abundance should be compared with regard to the common qualifications within each group. The aim of all such types of analysis is to be able always to

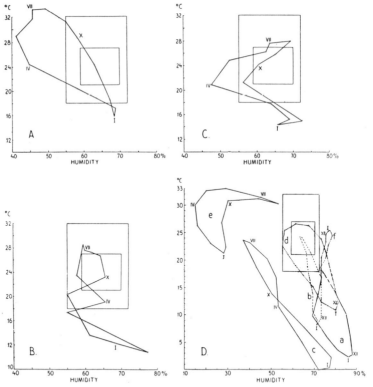

Fig. 9. Climograms of *Anychus latus*: A to C. in Palestine; A. at Jericho; B. at Dagania; C. at Rehobot; D. in various localities: a. at Paris; b. at Naples; c. at Ankara; d. at Cairo; e. at Khartoum; f. at Lourenço Marques.

circumscribe at least one climatically favourable zone. In the best case, we may be able to circumscribe an optimal, favourable, unfavourable, and impossible climatic zone.

Whenever this zonation is based on sufficiently large empirical data it becomes highly important for the understanding of seasonal fluctuation, local gradation, and geographical distribution. We shall give a few illustrations.

Fig. 9 shows regional climograms of the various citrusgrowing zones in Israel with regard to the Oriental Red Spider *Anychus latus*. The inner rectangle designates the optimal zone, the outer rectangle (broken line) the limit of the favourable zone for this mite. It is now obvious why, at Jericho, the fluctuation peaks are in autumn and late winter, at Dagania (Lake Tiberias) from April until November, and in the coastal plain (Rehoboth) from June to October. This seasonal fluctuation is in total agreement with that actually observed[13].

Diagram D in Fig. 9 explains clearly why six localities in six different zoogeographical regions do not offer an opportunity for any season gradation or high population level of *A. latus*, except to some degree at Cairo.

Whereas temperature always forms part of the climogram, it is not always clear whether humidity should be applied as precipitation or as relative humidity. This question may be settled only after knowing the biology of the species. Species, the eggs, larvae, and pupae of which dwell in the soil, as in the case of Noctuid moths, are subjected mainly to the influence of precipitation and the epidemics which they bring about. Flies, etc., which are mainly subjected to air-humidity, on which depends their adult longevity, are most suitably analysed with regard to air-humidity.

We mention here the classical research of W. C. Cook[22] concerning the climograms of some noctuid pests in North America. He discussed the consequences of weather distribution for zoogeographical as well as for epidemiological conclusions.

V. Shelford[23, 24] recently enlarged the use of the climogram to other than the hitherto employed factors, such as ultra-violet radiation, percentage of maximum possible sunshine, etc. He also made the first successful attempts to apply them to homoeotherms, using the most sensitive month of the year in particular. He described, e.g. the influence of weather on the populations of the Prairie Chicken *(Tympanuchus cupido)* in Illinois, as follows:

"Probably the most important critical period is the last phase of the growth of the gonads chiefly in April. It is evident that the combination at that time of 50 to 65% possible sunshine paired with 50 to 310 mm. of rain was associated with the highest number of booming males in the following year. However, many things probably happen to a population in the other eleven months. What should be known is the number of eggs produced and the relative vigour of the young."

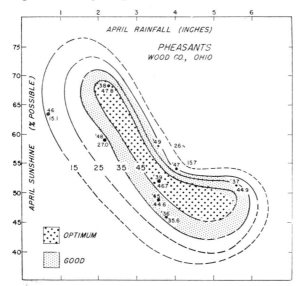

Fig. 10. A heliohydrogram, showing the number of pheasant nests per 100 acres as recorded in northwestern Ohio the year following the one plotted. The numbers are plotted at the intersections of coordinates presenting the amount of sunshine in April in precentage of the total possible (recorded at Fort Wayne, Ind.) and rainfall in April (recorded at Bowling Green, Ohio). The interaction of these factors appears to have had an important influence on the size of population. Although for best results these records should have been made in the study area, it is suggested that the same approximate number of nests occurs in series of different combinations of rainfall and sunshine; e.g. approximately 47 nests per 100 acres fell on approximately 68 % sunshine and 2.25 inches of rain and also on 53 % sunshine and 3.5 inches of rain; the data are too few but the ellipses shown follow the general pattern of such relationships. Optimum conditions, based on the largest number of nests, are in the center. The stippled areas indicate two zones of favorability in conditions, and the lines connect years of approximately equal population. Data are from E. Dustman's release Nr. 203 (1950) of the Ohio Wildlife Research Unit, Ohio State University. (From SHELFORD V.E., An Experimental Approach to the Study of Bird Populations. The Wilson Bulletin, Vol. 66, No. 4 (1954), pp. 253—258).

Incubation is also important in May, which seems, however, not to be a critical period with regard to the physical environment. June again, with hatching and rearing of the young, is critical. The paired temperature and rainfall appear important. The highest populations come about at 20.5 to 23.5° C and 64 to 117 mm rain.

A first tentative analysis of the Pronghorn Antelope *(Antelope cervicapra)* in the Yellowstone Park[25], where the peak of mating falls in October, puts the sensitive period of gonad development and reproductive stimulation in September (as in the Cottontail Rabbit). An optimal intensity of ultraviolet (i.e. another way of expressing percentage of possible sunshine) combined with suitable moisture results in more numerous and more vigorous offspring in a count eight months later, respectively 17 months after mating. The young are born in June and by September are half the weight of the adults.

H. WALLGREN[26] studied the climograms of the Yellow Bunting *(Emberiza citrinella)* and of the Ortolan Bunting *(E. hortulana)*. While the Yellow Bunting winters mainly within its breeding range, in Finland, with vagrating weather movements due to local changes in temperature and food-supply (after snowstorms), only a small proportion extends its vagrations beyond the south of the entire breeding area. The Ortolan Bunting, on the other hand, is a real migrant with well separated breeding and wintering areas. The comparison of the photoperiod-temperature climograms of both species in the breeding and wintering area show for the Yellow Bunting, of course, almost exclusive coverings in the climogram (Fig. 11), while the temperature conditions are similar for both seasons, the photoperiod being less favourable in winter. By migration very constant conditions of metabolism are secured from an energetical point of view. "The environmental factors most

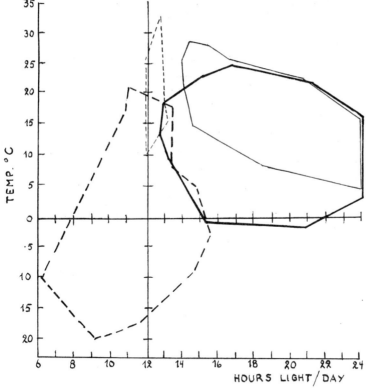

Fig. 11. Combined climograms of the migrating ortolan and the vagrating yellow bunting. Ortolan: full thin line- in summer; broken thin line- in winter; yellow buntings: full heavy line- in summer; broken heavy line- in winter. Abscissa: day length (hours light per day); ordinate: temperature. (After WALLGREN).

effective in stimulating northward migration in the Ortolan Bunting seem to be rising temperature and the beginning of the rainy season. In my opinion the species in its present physiological status would not be able to breed successfully within the wintering area (the High Yemen). It does not seem possible to demonstrate an environmental stimulus directly influencing the autumn migration of the ortolan. The present behaviour of the species is presumably the outcome of an evolutionary process in which temperature and photoperiod have been important factors." Birds migrating over the desert are in search of a suitable environment for metabolic equilibrium, recognised by the similarity of the photoperiod-temperature climogram. This is just another case where wisely chosen paired factors open up roads for the understanding of old biological problems.

WELLINGTON[27] has recently combined the analysis of macro- and micro-climate with that of outbreak history, behaviour and sensitivity of two Canadian Fire insects, *Choristoneura fumiferana* and *Malacosoma disstria* in a most promising way. Both species are active during the same season, both are adversely affected by prolonged rain at that period. *Choristoneura* develops best in dry, sunny weather, while *Malacosoma* develops better during moderately warm, humid, partly cloudy weather. Where both species coexist, outbreaks of them begin in different years, which in each case are preceded by a few years of specifically favourable weather. Before *Choristoneura* outbreaks northern and western air is dominating, whereas before *Malacosoma* outbreaks, it is more influenced by southern and southeastern air. *Choristoneura* outbreaks begin in suitable Boreal Fire areas two to four years after the continental complex of storm tracks has shifted northward, or after a marked increase in cyclonic activity. On the other hand, *Malacosoma* outbreaks tend to occur after southward shifts of the air tracks, or after decreased cyclonic activity. Continued studies of behaviour and sense physiology of both species in connection with the microclimate of the Fire forests give actual manifestation to different physical conditions.

Thus, new types of climograms come more to the foreground of ecological research. But the last two of our illustrations indicate clearly that they can be only fully evaluated when the animal's behaviour is taken fully into consideration, beyond the physical conditions.

### 4. The Epidemiological Bonitation of Ceratitis capitata

We shall now apply modern methods to the epidemiological bonitation of the Mediterranean fruit fly, *Ceratitis capitata*[28]. We may define bonitation as the estimate of numerical development of an animal at any given locality or season, and its menace of danger if it is an agricultural or medical pest.

The three hyperbolae for *Ceratitis* are:

|  | A | B | C |
|---|---|---|---|
| Threshold of development, °C | 12.4 | 12.4 | 12.4 |
| Thermal constant, days-degrees | 339 | 745 | 1,218 |
| Duration in days of development | | | |
| at 35° C | 15 | 33 | 54 |
| at 30° C | 19 | 43 | 60 |
| at 25° C | 27 | 59 | 97 |
| at 20° C | 45 | 98 | 160 |
| at 15° C | 130 | 287 | 469 |

Hyperbola B represents the epidemiological average of annual generations. But a few individual flies may survive for periods much longer than those given by hyperbola C, and those few survivors are of paramount importance for the maintenance of the species during unfavourable seasons, in which suitable hosts are rare or absent.

The physical zones of comfort of the species are:

| Zone | Temperature range in °C | Humidity range in % of R.H. | Bonitation index |
|---|---|---|---|
| Optimal | 16–32 | 72–85 | 2 |
| Favourable | 10–35 | 60–90 | 1 |
| Non-favourable | 2–38 | 40–100 | 0 |
| Impossible | < 2–40 < | < 40 | 0 |

In this way we obtain a reliable basis for climatic bonitation of *C. capitata*. The occurrence of 1–2 months, at least, within the impossible zone inhibits any permanent occurrence of the fly in the locality concerned, at least with regard to damage. In the non-favourable zone population density will remain very low. It will be higher in the favourable zone, and highest where favourable and optimal conditions exist for some consecutive months. In summing up the monthly values of the bonitation-index, we get the climatic local annual bonitation-value.

A series of additional corrections may be made to improve this bonitation (and others may be worked out locally):

(1) Rainfall of more than 125 mm combined with low temperatures (below 15° C) is detrimental, especially to the pupae in the soil and to the adult flies. Such months should be corrected by deducting one unit from the sum of positive values. Prolonged lack

of rains (irrigation lacking) is detrimental to the fly, directly by its influence on humidity and indirectly by its influence on vegetation.

(2) The biological unit for measuring the importance of any factor to which the fly is exposed is the generation. The speed of its development depends mainly on temperature, and a special indicator consigns this speed of development in each month, as shown opposite.

(3) Of greater importance, however, than both these factors are the host conditions. Abundant supply of favourable and suitable fruits for oviposition and development throughout the physically favourable season is optimal. Conditions are less favourable when the hosts are present during a shorter period only, or when they are unsuitable, i.e. if eggs or maggots show a high nutritional mortality during development. Conditions grow uncomfortable when the sequence of more or less suitable fruits is interrupted for

|  | 1 generation in | Generation factor |
|---|---|---|
| 10° C | ∞ days | 0 |
| 10–15° C | 50 days (no oviposition) | 1 |
| 15–25 and 32° C | 29–50 days (including preoviposition period) | 2 |
| 25–32° C | 18–29 days (including preoviposition period) | 3 |

a more or less prolonged season and when hosts grow scarce in number. Wherever these conditions become even more extreme, the host-factor is very unfavourable to impossible. The bonitation concerning the host-factor should never be made for a whole country or region, but must be determined for each locality separately. Considering the importance of its host-factor and its dependency on two major components, we express it as a fraction:

$$\frac{\text{host suitability}}{\text{host abundance}}$$

For the final damage bonitation we now proceed as follows: the weather and generation-factors are multiplied by the sum of both host-factors, and eventually – before multiplication – the unfavourable rain-factor is deducted from the climatic factor. In the case when the host-factor is zero, it should be multiplied by the generation factor only, only the climatic index being of importance.

The weather bonitation of 10 years for Tel-Aviv is:

| Year | Month | | | | | | | | | | | | In optimal | In favourable | In nonfavourable | In impossible | Total |
|---|---|---|---|---|---|---|---|---|---|---|---|---|---|---|---|---|---|
| | i | ii | iii | iv | v | vi | vii | viii | ix | x | xi | xii | | | | | |
| 1927 | 1 | 1 | 2 | 2 | 2 | 2 | 2 | 2 | 2 | 2 | 2 | 1 | 9 | 3 | .. | .. | 21 |
| 1928 | 1 | 1 | 1 | 1 | 1 | 2 | 2 | 2 | 2 | 2 | 1 | 1 | 5 | 7 | .. | .. | 17 |
| 1929 | 1 | 1 | 1 | 2 | 1 | 2 | 2 | 2 | 1 | 1 | 2 | 1 | 5 | 7 | .. | .. | 17 |
| 1930 | 1 | 1 | 1 | 1 | 1 | 1 | 1 | 1 | 1 | 1 | 1 | 1 | .. | 12 | .. | .. | 12 |
| 1931 | 1 | 1 | 1 | 1 | 1 | 1 | 1 | 1 | 1 | 1 | 1 | 1 | 1 | 11 | .. | .. | 13 |
| 1932 | 1 | 1 | 1 | 1 | 1 | 1 | 1 | 1 | 1 | 1 | 1 | 0 | .. | 11 | 1 | .. | 11 |
| 1933 | 1 | 1 | 0 | 1 | 1 | 1 | 2 | 2 | 2 | 2 | 1 | 1 | 4 | 7 | 1 | .. | 15 |
| 1934 | 1 | 1 | 1 | 2 | 2 | 2 | 2 | 2 | 1 | 2 | 1 | 1 | 6 | 6 | .. | .. | 18 |
| 1935 | 1 | 1 | 1 | 1 | 1 | 2 | 2 | 2 | 1 | 2 | 2 | 1 | 5 | 7 | .. | .. | 17 |
| 1936 | 1 | 1 | 1 | 1 | 2 | 2 | 2 | 1 | 1 | 1 | 2 | 1 | 4 | 8 | .. | .. | 16 |

The climogram of *Ceratitis* at Tel-Aviv (Fig. 12) confirms this important fluctuation, which for Jerusalem is still more important (minimum 1 to maximum 8) whereas in Jericho conditions are unfavourable (index: 0 to 2).

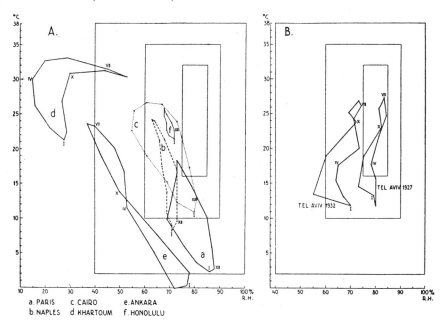

Fig. 12. Climograms of *Ceratitis capitata*. A. in various regions: a. at Paris; b. at Naples; c. at Cairo; d. at Khartoum; e. at Ankara; f. at Honolulu. B. at Tel Aviv in the fruitfly year 1927 and in 1932 when weather did not favour it.

A complete bonitation of an orange grove in the neighbourhood of Tel Aviv in 1932 and 1934 respectively runs as follows:

**Table VIII.**
Complete *Ceratitis*-bonitation of an Orange Grove in the Neighbourhood of Tel Aviv in 1932 and 1934

| | i | ii | iii | iv | v | vi | vii | viii | ix | x | xi | xii | |
|---|---|---|---|---|---|---|---|---|---|---|---|---|---|
| **1932** Weather | 11.9 | 13.2 | 15.5 | 17.7 | 20.1 | 24.2 | 26.2 | 26.9 | 25.1 | 23.8 | 19.0 | 13.5 | °C |
| | 70 | 67 | 64 | 65 | 73 | 71 | 74 | 73 | 69 | 72 | 60 | 55 | % R.H. |
| | 131 | 96 | 8 | 3 | .. | .. | .. | .. | 1 | 38 | 58 | 43 | mm. rains |
| Bonitation | 1 | 1 | 1 | 1 | 1 | 1 | 1 | 1 | 1 | 1 | 1 | 0 | Climatic index: 11 Rain-factor |
| | −1 | .1 | .2 | .2 | .2 | .2 | .3 | .3 | .3 | .2 | .2 | .1 | Generation-factor |
| | 1/3 | 1/3 | 1/3 | 0 | 0 | 0 | 0 | 0 | 1/3 | 1/3 | 1/3 | 1/3 | Host-factor |
| **Result 1932** | 4 | 4 | 8 | 1 | 1 | 1 | 1 | 1 | 12 | 8 | 8 | 0 | 49 |
| **1934** Weather | 12.8 | 11.7 | 16.3 | 18.5 | 21.6 | 24.9 | 26.7 | 27.1 | 25.7 | 23.2 | 19.2 | 14.4 | °C |
| | 76 | 79 | 71 | 80 | 80 | 77 | 79 | 79 | 74 | 75 | 73 | 80 | % R.H. |
| | 165 | 152 | 10 | 11 | 1 | .. | .. | .. | .. | 1 | 13 | 445 | mm. rains |
| Bonitation | −1 | −1 | 1 | 2 | 2 | .2 | .2 | .2 | .1 | 2 | 1 | −1 | Climatic index: 18 Rain-factor |
| | −1 | .. | .2 | .2 | .2 | .2 | .3 | .3 | .3 | .2 | .2 | .1 | Generation-factor |
| | 1/3 | 1/3 | 1/3 | 0 | 0 | 0 | 0 | 0 | 1/3 | 1/3 | 1/3 | 1/3 | Host-factor |
| **Result 1934** | 4 | 4 | 8 | 2 | 2 | 2 | 2 | 2 | 12 | 16 | 8 | 4 | 66 |
| Changed host-factor | 1/3 | 1/3 | 1/3 | 2/2 | 2/2 | 2/2 | 2/2 | 2/2 | 1/3 | 1/3 | 1/3 | 1/3 | |
| Result: 1932 | 4 | 4 | 8 | 8 | 8 | 8 | 12 | 12 | 12 | 8 | 8 | 0 | 92 |
| Result: 1934 | 4 | 4 | 8 | 16 | 16 | 16 | 24 | 24 | 12 | 16 | 8 | 4 | 152 |

In Tel Aviv climatic conditions are favourable to optimal all the year round. In 1932 they were favourable, except for December. The rain-factor is only slightly unfavourable in January. No extreme drought decimates the adult fly population seriously. The generation factor is very favourable from March to November. Host-sequence is fairly favourable, but is interrupted for 5 months. The main host is not very suitable. No high fly population can be present in September, but damage of late fruit may be expected from late March on.

In 1934 conditions are rather different. Whereas host- and generation-factor remain constant, the rain-factor grows more unfavourable in December. The high precipitation in December has an effect only in the following year. Weather conditions, however, are far more favourable. They are optimal from April to August, very close to optimal in September, and optimal again in October. No interruption occurs in December. The optimal climatic conditions in spring and summer permit the survival of a relatively high number of flies, which should be able and actually has been able to attack citrus fruits in September to a considerable extent. These conditions are improved, as far as the fly is concerned, by the fact that some other fruits are almost always available within the garden or in its neighbourhood throughout spring and summer. Apricots, prickly pear, figs, *Lyctus*, etc., are among the more common ones. The quantity and distance of these intermediate fruits determine the intensity of the attack on oranges in autumn. An uninterrupted sequence of suitable hosts, as has been alternatively calculated in the preceding table, would have resulted in serious damage in 1932, and in extremely serious damage in 1934.

We must content ourselves with regard to the value of the weather bonitation factor in other localities within the area of distribution of *C. capitata*. Table IX serves as a starting-point in this discussion.

In the humid tropics favourable conditions for *Ceratitis* prevail everywhere in the lowlands, and at moderate elevations. This zone is probably the original home of the Mediterranean fruit-fly. In deserts (Sahara-Sindian and Northern Sudanian regions) conditions are extremely unfavourable if special conditions do not maintain a high average air-humidity, as in Cairo. The same is true for steppe habitats (Irano-Turanian region). Mediterranean types of climate are rather suitable to the development of *Ceratitis*, as long as a relatively high humidity is maintained during the summer. The warmer parts of the region, especially, show high bonitation values, but lower ones than in the humid tropics. Winter rains and winter cold are rarely severe and long enough to exterminate the adult population.

The careful observer of the table will remark that the number of possible annual generations (hyperbola B) gives reliable results

## Table IX.
Climatic bonitation of some localities for *Ceratitis capitata*

| Locality | No. of possible generations | | | No. of months in zone | | | No. of months below 12.4° C | Climatic bonitation | Region |
|---|---|---|---|---|---|---|---|---|---|
|  | A | B | C | Opt. | Fav. | N. fav. | Imposs. |  |  |  |

| Locality | A | B | C | Opt. | Fav. | N. fav. | Imposs. | No. of months below 12.4° C | Climatic bonitation | Region |
|---|---|---|---|---|---|---|---|---|---|---|
| Berlin | 1.8 | 1.1 | 1.0 | . | . | 5 | 4 | 3 | 7 | 0 | Euro-Siberian |
| Paris | 1.6 | 1.1 | 0.9 | . | . | 6 | 6 | . | 7 | 6 | Euro-Siberian |
| Pinsk | 1.7 | 1.1 | 0.9 | . | . | 5 | 2 | 5 | 7 | 0 | Euro-Siberian |
| Ankara | 3.7 | 1.5 | 1.3 | . | . | . | 9 | 3 | 6 | 0 | Irano-Turanian |
| Damascus | 6.3 | 3.2 | 2.2 | . | . | 1 | 9 | 2 | 4 | 1 | Irano-Turanian |
| Nice | 3.3 | 1.7 | 1.4 | . | . | 7 | 5 | . | 6 | 7 | Mediterranean |
| Naples | 4.8 | 2.5 | 1.9 | . | . | 9 | 3 | . | 4 | 9 | Mediterranean |
| Malaga | 6.6 | 3.5 | 2.5 | . | . | 12 | . | . | 1 | 12 | Mediterranean |
| Tel Aviv | 8.2 | 4.2 | 2.8 | 2 | 10 | . | . | . | . | 14 | Saharo-Sindian |
| Cairo | 7.6 | 3.7 | 2.6 | 2 | 8 | 2 | . | . | 2 | 12 | Saharo-Sindian |
| Bagdad | 11.0 | 5.4 | 3.2 | . | 4 | 5 | 3 | . | 2 | 0 | Sudanian |
| Khartoum | 16.7 | 7.4 | 4.3 | . | . | 3 | 9 | . | . | 0 | Sudanian |
| Mongalla | 14.6 | 6.7 | 3.9 | 6 | 3 | 3 | . | . | . | 15 | Sudanian |
| Lourenço Marques | 10.2 | 4.9 | 3.1 | 7 | 5 | . | . | . | . | 19 | Ethiopian Savana |
| Salisbury | 6.6 | 3.5 | 2.5 | 4 | 4 | 4 | . | . | 4 | 12 | S. Africa |
| (Batavia) | 14.8 | 6.6 | . | 12 | . | . | . | . | . | 24 | Oriental Region |
| (Manila) | 15.1 | 4.7 | . | 8 | 4 | . | . | . | . | 20 | Oriental Region |
| Honolulu | 11.6 | 5.5 | 3.2 | . | 12 | . | . | . | . | 12 | Hawaii |
| Tampa | 10.4 | 4.7 | 2.9 | 2 | 10 | . | . | . | . | 14 | (near) Florida |
| San Francisco | 1.6 | 0.7 | 0.5 | . | 12 | . | . | . | 4 | 12 | California |
| Los Angeles | 4.9 | 2.2 | 1.4 | . | 9 | 3 | . | . | . | 9 | California |
| San Diego | 4.0 | 1.8 | 1.2 | 4 | 8 | . | . | . | . | 16 | California |
| Fresno | 6.1 | 2.7 | 1.6 | . | 4 | 4 | 4 | . | 3 | 0 | California |

only in combination with the climatic bonitation. However, together they seem to give a fairly reliable picture of the climatic potential existing at any given locality for *C. capitata*.

We shall glance at the bonitation obtained for U.S.A. Hawaii, with a factor of 12 and 5.5 annual generations, as well as Tampa, in the centre of the now exterminated heavy infestation in Florida, with 4.7 generations and a factor of 14–17, offer decidedly optimal conditions throughout the year. At San Francisco all months are in the favourable zone, but fall below the development threshold in 4 of the temperatures and only 0.7 generations per annum may develop. During 4 months even optimal conditions prevail. In Los Angeles 9 months are in the favourable and 3, but these the less important winter months, are in the unfavourable zone, owing to low humidity. The latter, in addition, is probably efficiently counterbalanced by the not too heavy winter rains. The number of 2.2 annual generations approaches the danger-zone. Some damage could certainly be expected at San Diego as well as at Los Angeles. In the dry interior of California the bonitation is very low. At Fresno the factor is depressed to zero owing to a sequence of 4 months of extremely low humidity.

Further co-operation is invited with regard to *C. capitata* in order to show the degree of agreement and to improve the bonitation by additional research.

## 5. Sense Ecology and Behaviour

The living organism, its behaviour, and its reactions under natural conditions have been neglected all too long in research, whereas in reality this knowledge is just as indispensable as the knowledge of physical and biotic factors of the environment[29].

Behaviour is the directive agent through the manifold and ever-changing complexities of the outer world (Fig. 13). It is by far the most important and oft-neglected guide to the animal through all ecological environmental factors. If, for example, a bloodsucking mosquito, gnat, or fly approaches man or another mammal, it is attracted by the smell of its sweat and cannot possibly have a clear conception that this sweat will lead it to the desired blood. Yet actually it does: – sucking on the sweat-secreting integument, it is led to the capillaries, the source of its food. We are unable to comprehend how the fly could apperceive directly and approach its food. By and by it will learn that the sweat leads to blood but this is a secondary phenomenon, if it occurs at all. In other cases we can exclude learning with certainty. In *Schistocerca gregaria* we observed that eggs are laid, as a rule, only in wet sand, and it is only in wet, well-aerated soil that the egg of the Desert locust can begin its development successfully[30]. The egg-laying female has, of course, not the slightest intention or understanding of this interrelation.

Still more complicated is the situation with regard to the Moroccan locust *(Dociostaurus maroccanus)*[31] which oviposits in dry soil. In wet soil the very young eggs would absorb water and die within one week, as is the case when an unexpected rain falls on the steppe in early June. Yet it also lays its eggs in bare patches of the soil surface of the steppe, in this region precisely where next spring – just when the young larvae will be hatching – the grass *Psoa bulbosa* (or a related congener) will develop, offering the soil-covering rosette of its lower leaves, the preferred food of the neonate *Dociostaurus* during the very first one or two days. The ovipositing female, here too, cannot possibly have any understanding of these interrelations, separated by a period of almost eleven months. Yet such almost miraculous interrelations are actually the anchor by which the animal is fixed in this or that environment and which enables it to survive. Often the number of such "chains" is quite numerous for the student who knows how to "wonder".

The behaviour of every species is, therefore, as integral a part of ecology as are the chemo-physical or the other biotic factors of its environment.

An inherited chain of reflexes connects the exigencies of egg-development with the behaviour of the female rather closely. A large number of similar cases is known, especially in connexion with safeguarding the early development of the offspring. The colours to which the hawk-moth *Macroglossa stellatarum* is attracted change, in the gravid female, to that of chlorophyll, and this attraction, coupled with the smell of etheric oils of *Galium*, guarantees the proper oviposition and development of the caterpillars[32].

Animals often migrate within their habitat into different strata or niches, or other parts of the same host (interior or exterior, crown or root). It is of the utmost importance to verify in such cases whether these migrations are actively provoked by attraction to the new micro-habitat or whether they are simply the result of selection and survival.

Taxes are often influenced by environmental factors. WEBER[33] has demonstrated in *Trialeurodes vapariorum* the following changes of behaviour in varying temperatures:

| Temperature, °C | 10–15 | 15–20 | 20–25 | 25–30 | 30–35 | 35–40 | 40–45 |
|---|---|---|---|---|---|---|---|
| Phototaxis | 0 | + | + | + | ++ | ++ | – |
| Geotaxis | 0 | ++ | ++ | + | + | – | – |

Success in finding food depends entirely upon the behaviour of the organism. It is necessary to know how far it may be able to direct itself actively towards its food, if at all, and if so, by which qualities of the food it is guided (chemical or optical, sense, etc.).

KENNEDY[34] has shown that the difference of one tactic reaction as compared in congeneric species (Dragonflies and Ephemerids) is sufficient to cause entrance into new niches or even new areas of distribution.

It is very regrettable that the field opened by the ingenious work of v. UEXKUELL[35] is worked upon so little. Recently work was begun to explore the problem of recognition of the sex-partner, of the fellow or of the enemy. Quality after quality is deducted until the proper recognition as sex-partner, etc. ceases.

VON UEXKUELL leads us to a first understanding of the different "worlds" or moods of any animal, which vary in dominance in alternation in the same individual. Now the chirping love song, then copulation followed by oviposition, then feeding, then migrating,

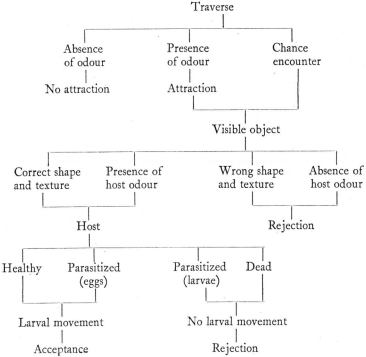

Fig. 13. Host selection by the hymenopterous wasp *Microplectron fuscipennis*. Tabular presentation of the course of events in the field (After ULLYETT).

followed by resting, and in the morning again sun barking followed by chirping, ever more synchronized: – these are the normal "worlds" or their expressions dominating all others, one after the other, in a Moroccan locust's mind on a normal oviposition day in May on the

Iraqui steppe[35]. Other worlds, like the preparedness to fight or to flee an enemy, etc. may be temporarily suppressed by the dominance, e.g. of the ovipositing world, certainly not to the good of the animal. HOWARD[36] had beautifully described a waterhen's *(Fulica atra)* worlds in spring. It would obviously be amiss to appreciate only the behaviour of a species, its worlds, etc. to the neglect of the physical environment.

The parasite *Microplectron fuscipennis* is able to discriminate easily between true and false hosts even where the latter resemble the normal host in everything except the presence of the living larva within the cocoon[37]. Acceptance of a host seems to depend, to a large extent, upon the presence of larval movement.

Great efficiency is shown in discriminating between healthy material and hosts containing parasite larvae which are well grown. The presence of parasite eggs, on the other hand, does not deter females from ovipositing in the host. The reason for this difference is either the larval movement or smell which is present in the latter and absent in the former case. There is thus a definite proportion of every host population which is not subject to random oviposition. A wholly mechanistic view of host selection is untenable, and the underlying basis of behaviour is of a psychological nature.

The presence or absence of social appetite[38] and its form and constancy, the form and constancy of sexual attraction, seasonal agglomerations or migrations, and their physiological and behaviour basis are of primary importance. The field and laboratory methods for studying those problems are rather inadequate today. It will be impossible, however, to work out standard methods, because every animal will require special methods to suit its special organization, requirements, and environment.

# III
# APPROACH TO AN ANALYSIS OF SOME ANIMAL POPULATIONS

## 1. *Drosophila* and the Logistic Curve

*Drosophila* has been bred for the last two decades in a great number of biological laboratories, mainly for genetical purposes. It was R. PEARL[1] who laid the foundations of an experimental study of animal populations and their analysis by breedings of *Drosophila melanogaster*. If one pair or a small population of *Drosophila* is introduced into milk bottles which are filled to a certain height with a standard larval food to which yeast is added, the following facts are always observed: The growth of population begins at a slow rate, then becomes rapid and successively slackens its speed again until a saturation point is reached at which no further increase of population occurs. The growth follows a sigmoid curve, called the logistic curve, and is observed again and again wherever we have an opportunity to study the growth of animal populations, be it that of Protozoa in the test-tube or that of fish in ponds or that of human populations. If environmental conditions are constant and the experiments are performed with a genetically uniform stock, the course of the curve is very uniform in parallel breedings.

The mathematical characteristics of the logistic curve are a lower and upper asymptote and a turning-point. Its general formula is

$$y = \frac{k}{1 + e^{a-bx}}$$

where $y$ is the size of the population, $x$ the time (in age of population), $k$ the upper asymptote, and $a$ and $b$ constants.

PEARL counted the adult fly population and its growth only, since he regarded all previous stages, from egg to pupa, as comparable with the embryonic development of mammals or birds. He attempted to approach the physiological basis of the dynamics of the logistic growth-curve by a series of laborious as well as ingenious experiments. In each of these experiments all factors remained apparently constant except one. This variable factor was, in some series, environmental (temperature, size or volume of milk bottle, composition of food, etc.), in other series genetical (wild strains and mutants of known genetical composition), while in a third series population density was the variable factor (longevity, fertility). It is to PEARL's credit to have discovered the basic ecological importance of the density-factor in population-growth and in animal populations in general.

The results obtained with regard to longevity and to genetical composition of the fly population have been made use of in others of these essays. A few further results may be given here.

Wild *Drosophila* bred under various conditions gave the following results[2]:

| Condition | Mean duration of life in days | No. of eggs per female-day | Total eggs per female | Wing length as indicator of body size |
|---|---|---|---|---|
| | | | | mm. |
| At 27° C | 27.7 | 29.3 | 812 | .. |
| Ditto, underfed | 39.1 | 9.9 | 387 | 2.81 |
| At 30° C . . . | 37.1 | 14.7 | 545 | 3.80 |
| In cold . . . . . | 31.0 | 27.5 | 853 | 2.53 |

We used a wild strain, the genetical constitution of which is entirely unknown, with which we have been experimenting for three consecutive years. The results seem consistent enough to justify the assumption that the strain is of sufficient genetical homogeneity to permit a population analysis. The volume of our milk bottles was 190 cc of which 50 cc were filled with agar 3 cm high. The agar used was the synthetic medium of PEARL (5–101). All populations were bred at 27° C.

Development in relation to temperature was observed at 15, 20, 24, and 27° C., from hatching of adults to the hatching of the first individuals of the filial generation. The hyperbola of development (including the pre-oviposition period of the adult) shows the following constants:

Threshold of development $(c)$ = 10.1 C.
Thermal constant (th.c.) = 152 days-degrees

The influence of fly-density on fertility (Fig. 14, B) was studied in two series of experiments in which, at 27° C, flies were transferred into new milk bottles for five successive days and the offspring bred:

| Density of adult flies | 1 | 5 | 10 | 20 | 40 | pairs |
|---|---|---|---|---|---|---|
| Total egg-production . . . . . | 409 | 656 | 553 | 599 | 1,031 | flies |
| Egg-production per female-day . . . | 45.4 | 14.7 | 6.1 | 3.3 | 2.8 | ,, |

Normal fertility of *Drosophila* females (Fig. 14, C) was studied in milk bottles with the synthetic food-medium, two pairs of flies per bottle.

The average longevity was 13.6 days, the average fertility 300.6 eggs per female. The values for the individual female-day averages fluctuated from 14.3 to 29.0 eggs, for total oviposition from 136 to 520 eggs per female.

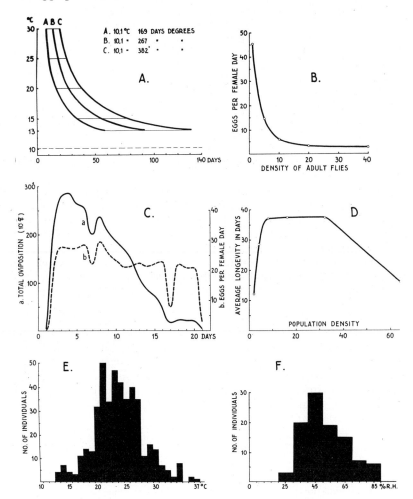

Fig. 14. Physical Ecology of *Drosophila melanogaster*. A. The three hyperbolae showing temperature-dependency of development (including pre-oviposition period), epidemiological average duration and longevity at various temperatures. B. Oviposition at different population densities. C. Total daily oviposition (a) and daily oviposition per female (b) of ten females. D. Adult longevity at different fly densities. E. Temperature preference of adult flies. F. Humidity preference of adult flies.

Fig. 14 A–E gives the results of a number of experiments with regard to physical ecology.

The scale of activity in 104 individual experiments was:

|     |                                                              | °C       |          |
| --- | ------------------------------------------------------------ | -------- | -------- |
| (1) | Upper limit of cold-torpor                                   | 6.1      | (5–7)    |
| (2) | Upper limit of uncoordinated occasional movements of extremities only | 11.0 | (10–12) |
| (3) | Upper limit of slow crawling                                 | 12.8     | (12–14)  |
| (4) | Normal activity                                              | 12.8–27.9 |         |
| (5) | Lower limit of high activity                                 | 27.9     | (27–29)  |
| (6) | Lower limit of excited activity                              | 32.2     | (27–34)  |
| (7) | Beginning of heat paralysis                                  | 37.8     | (34–39)  |
| (8) | Instantaneous heat-death                                     | 39.6     | (38–41)  |

The normal range of activity extends from 12.8 to 27.9° C; the high activity from 27.9 to 32.2° C is certainly above the comfort-, but still below the danger-zone.

In the normal culture the number of adults begins to increase on the 10th day. The population maximum of 230 adult flies is reached on the 23rd day and is maintained with small fluctuations until the 31st day. The population then decreases until it is almost zero on the 50th day. The agar has dwindled away to a thin hard mass by this time and is entirely unfit as physical environment for the larvae. We had little doubt that if this food-medium could be kept fresh, a normal fluctuating population could be maintained for a long time.

Since PEARL compares the growth of human populations to that of *Drosophila*, it seemed to be of interest to see how artificial immigration of adult fly populations in the early part of population growth would influence the environmental capacity, i.e. the upper asymptote. Two series of observations were made, one with 6 flies introduced 5 times, from the 3rd to the 8th day, another with 30 flies introduced on the 3rd day of the culture. The results ($k = 253$,

### Description of Cultures of Table X

1. Normal population growth in milk bottle of 190 c.c. volume.
2. Normal population growth in milk bottle of 190 c.c. volume.
3. Normal population growth in milk bottle of 190 c.c. volume.
4. Normal population growth (average of 1-3).
5. Normal population growth with immigration of 6 adult flies from the 3rd to 8th day.
6. Normal population growth with immigration of 30 adult flies on the 3rd day.
7. Population growth in cylinder with 26.4 c.c. agar -149.6 c.c. airspace.
8. Population growth in cylinder with 26.4 c.c. agar -74.8 c.c. airspace.
9. Population growth in cylinder with 52.8 c.c. agar -74.8 c.c. airspace.
10. Population growth in Erlenmeyer flask with 85 c.c. agar 40 c.c. air-space.
11. Population growth in Erlenmeyer flask with 122 c.c. agar 128 c.c. air-space.
12. Population growth in Erlenmeyer flask with 255 c.c. agar 65 c.c. air-space.
13. Population growth in Erlenmeyer flask with 350 c.c. agar 65 c.c. air-space.

71

**Table X.**
Population growth of our strain of *Drosophila* (each series being the average of 2 – 3 breedings)

| Day | 1 | 2 | 3 | 4 | 5 | 6 | 7 | 8 | 9 | 10 | 11 | 12 | 13 |
|---|---|---|---|---|---|---|---|---|---|---|---|---|---|
| 1 | 4 | 4 | 4 | 4 | 4 | 4 | 4 | 4 | 4 | 4 | 4 | 4 | 4 |
| 2 | 4 | 4 | 4 | 4 | 4 | 4 | 4 | 4 | 4 | 4 | 4 | 4 | 4 |
| 3 | 4 | 4 | 4 | 4 | 10 | 34 | 4 | 4 | 4 | 4 | 4 | 4 | 4 |
| 4 | 4 | 4 | 4 | 4 | 16 | 34 | 4 | 4 | 4 | 4 | 4 | 4 | 4 |
| 5 | 4 | 4 | 4 | 4 | 22 | 34 | 4 | 4 | 4 | 4 | 4 | 4 | 4 |
| 6 | 4 | 4 | 4 | 4 | 28 | 34 | 4 | 4 | 4 | 4 | 4 | 4 | 4 |
| 7 | 4 | 4 | 4 | 4 | 34 | 34 | 4 | 4 | 4 | 4 | 4 | 4 | 4 |
| 8 | 4 | 4 | 4 | 4 | 34 | 34 | 4 | 4 | 4 | 4 | 4 | 4 | 4 |
| 9 | 4 | 4 | 4 | 4 | 34 | 34 | 4 | 4 | 4 | 4 | 4 | 4 | 4 |
| 10 | 14 | 23 | 34 | 24 | 34 | 52 | 4 | 4 | 4 | 6 | 7 | 6 | 6 |
| 11 | .. | .. | .. | .. | 81 | .. | 54 | 104 | 73 | 31 | 39 | 28 | 20 |
| 12 | 110 | 106 | 118 | 111 | .. | 100 | 120 | 193 | 134 | 69 | 81 | .. | 44 |
| 13 | 148 | 133 | 176 | 152 | 126 | 134 | 190 | 205 | 160 | 84 | 102 | 132 | 82 |
| 14 | 116 | 152 | 183 | 167 | 111 | 145 | 206 | 203 | 165 | 125 | 176 | 236 | .. |
| 15 | 176 | 167 | 225 | 189 | 161 | 149 | 198 | 162 | 154 | 146 | 217 | 292 | 131 |
| 16 | 189 | 169 | 205 | 188 | 158 | 173 | 120 | 203 | 160 | 139 | 190 | 307 | 149 |
| 17 | 188 | 234 | 174 | 199 | 181 | 193 | 139 | 101 | 189 | 109 | 167 | 390 | 110 |
| 18 | 217 | 181 | 235 | 211 | 184 | 209 | 109 | 54 | 139 | 122 | 117 | 340 | 36 |
| 19 | 204 | 185 | 230 | 206 | 195 | 203 | 66 | 64 | 129 | 94 | 84 | 44 | 5 |
| 20 | 232 | 180 | 235 | 216 | 184 | 209 | 62 | 56 | 123 | 111 | 108 | 443 | 10 |
| 21 | 234 | 193 | 243 | 223 | 204 | 232 | 50 | 85 | 145 | 112 | 120 | 354 | .. |
| 22 | 226 | 189 | 254 | 223 | 201 | 231 | 30 | 99 | 115 | 176 | 182 | .. | 215 |
| 23 | 209 | 204 | 278 | 230 | 211 | 244 | 45 | 83 | 141 | 179 | 212 | 368 | 525 |
| 24 | 199 | 186 | 208 | 198 | 222 | 244 | .. | .. | .. | 218 | 218 | 308 | |
| 25 | 178 | 211 | 265 | 218 | 214 | 246 | 66 | 118 | 205 | 184 | 152 | 259 | a |
| 26 | 177 | 191 | 266 | 210 | 253 | 236 | 62 | 88 | 198 | 125 | 117 | 320 | g |
| 27 | 178 | 202 | 253 | 211 | 248 | 232 | 85 | 105 | 172 | .. | .. | 280 | a |
| 28 | 179 | 202 | 283 | 221 | 239 | 197 | 101 | 113 | 128 | 55 | 96 | 230 | r |
| 29 | 207 | 200 | 234 | 214 | 224 | 188 | 107 | 86 | 86 | 32 | 54 | 228 | .. |
| 30 | 224 | 182 | 235 | 214 | 224 | 176 | 91 | 77 | 83 | 91 | 65 | 226 | l |
| 31 | 232 | 201 | 230 | 221 | 214 | 183 | .. | .. | .. | 30 | 51 | 200 | o |
| 32 | 156 | 232 | 140 | 176 | 225 | .. | 77 | 62 | 63 | .. | .. | 157 | o |
| 33 | 79 | 245 | 78 | 134 | .. | 170 | 35 | 45 | 51 | 29 | 123 | 120 | s |
| 34 | 78 | 226 | 86 | 152 | 198 | 197 | 11 | 18 | 57 | 37 | 130 | 115 | e |
| 35 | 76 | 254 | 67 | 132 | 187 | 189 | .. | .. | .. | .. | 91 | 102 | .. |
| 36 | 73 | 287 | 61 | 140 | 171 | 178 | .. | .. | .. | 35 | 70 | 81 | .. |
| 37 | 75 | 178 | 72 | 108 | 141 | 159 | .. | .. | .. | 36 | .. | 72 | .. |
| 38 | .. | .. | .. | .. | 107 | 137 | .. | .. | .. | .. | .. | 101 | .. |
| 39 | 69 | 99 | 77 | 82 | 93 | .. | .. | .. | .. | .. | 76 | 100 | .. |
| 40 | 66 | 88 | 75 | 76 | .. | 99 | .. | .. | .. | .. | .. | 125 | .. |
| 41 | .. | .. | .. | .. | 71 | 89 | .. | .. | .. | .. | .. | .. | .. |
| 42 | 68 | 74 | 37 | 60 | 72 | .. | .. | .. | .. | .. | .. | .. | .. |
| 43 | 62 | 82 | 17 | 54 | 69 | 83 | .. | .. | .. | .. | .. | .. | .. |
| 44 | 46 | 60 | 5 | 37 | 58 | 88 | .. | .. | .. | .. | .. | .. | .. |
| 45 | 37 | 57 | 2 | 32 | .. | .. | .. | .. | .. | .. | .. | .. | .. |
| 46 | 32 | 41 | .. | .. | .. | .. | .. | .. | .. | .. | .. | .. | .. |
| 47 | 23 | 40 | .. | .. | .. | .. | .. | .. | .. | .. | .. | .. | .. |
| 48 | 17 | 25 | .. | .. | .. | .. | .. | .. | .. | .. | .. | .. | .. |
| 49 | 5 | 10 | .. | 5 | .. | .. | .. | .. | .. | .. | .. | .. | .. |
| 50 | .. | .. | 3 | 1 | .. | .. | .. | .. | .. | .. | .. | .. | .. |

246, and 230 = 243) are identical with those of the normal individual cultures ($k = 234$, 234, and $283 = 250$), as was to be expected.

The formula of the logistic curve (as based on the average in No. 4) is for the normal cultures:

$$y = \frac{230}{1 + e^{1.9299 - 0.1702x}}$$

The analytical study of the dynamics of bee populations made it clear that no dynamic analysis of population trends can be attempted without counting all stages of the species concerned. We have criticized before, from another point of view, the omission of any age from general life-tables or other statistical population census. It was attempted, therefore, to perform complete daily counts of *Drosophila* of all ages from egg to adult.

A culture cannot be continued after it has been counted, as the entire agar is taken out and carefully sliced in order to get a complete census. The experiments were, therefore, arranged by a simultaneous preparation of 90 bottles, each of which was colonized on the same morning by 2 pairs of flies hatched during the 2 hours preceding the transfer. Two bottles were counted and destroyed daily. The course of the experiments demonstrated that quite consistent results were obtained in this way, forming a reliable basis for population analysis. The differences do not surpass those on which STANLEY based his complicated and successful mathematical analysis of *Tribolium*-populations, but are rather smaller[3].

We shall first condense in Table XI the results of 4 different series of daily population counts from the beginning to the 32nd day of age. Two series only are counted from the 22nd day on, as the other two breedings were destroyed after that day by mites which fed on the *Drosophila* eggs (Fig. 15, A)[4].

The formula of the logistic curve is:

$$y = \frac{719}{1 + e^{2.3877 - 0.3591}}$$

Potential development, when initial conditions continue to prevail, needs little comment. It exposes the typical exponential curve of unlimited growth.

The table in which all eggs are assumed to develop into flies is of interest because it shows that even with a steadily increasing number of adult flies oviposition would cease by force of density-pressure alone.

The data concerning the change of adult physiology with progression of age of the *Drosophila* culture are also shown in Fig. 16.

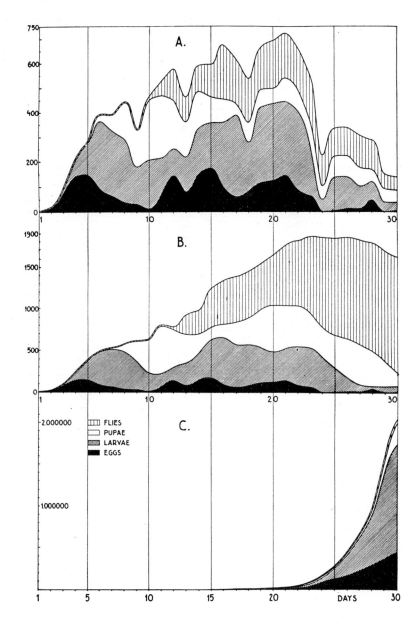

Fig. 15. Actual and potential growth in the *Drosophila* bottle. A. Actual observations. B. If all eggs actually laid would develop into adult flies. C. If the population would increase at the intensity of the initial growth (13 days adult longevity and 300 eggs per female).

**Table XI.**

Average of four series of breedings of *Drosophila* (from 22nd day two series only)

| Day | Eggs | Neonate larvae | Total larvae | Pupae | Flies living | Dead Flies | Total |
|---|---|---|---|---|---|---|---|
| 1 | .. | .. | .. | .. | 4 | .. | 4 |
| 2 | 16 | .. | .. | .. | 4 | .. | 20 |
| 3 | 78 | 10 | 10 | .. | 4 | .. | 92 |
| 4 | 144 | 57 | 65 | .. | 4 | .. | 213 |
| 5 | 148 | 85 | 134 | .. | 4 | .. | 268 |
| 6 | 290 | 91 | 257 | 25 | 4 | .. | 378 |
| 7 | 65 | 62 | 264 | 66 | 4 | .. | 399 |
| 8 | 38 | 87 | 255 | 150 | 4 | .. | 447 |
| 9 | 34 | 51 | 152 | 149 | 4 | .. | 339 |
| 10 | 12 | 58 | 199 | 242 | 15 | .. | 468 |
| 11 | 76 | 19 | 143 | 250 | 61 | .. | 530 |
| 12 | 147 | 40 | 109 | 196 | 127 | 3 | 579 |
| 13 | 83 | 37 | 141 | 139 | 200 | 4 | 563 |
| 14 | 153 | 61 | 181 | 152 | 204 | 5 | 690 |
| 15 | 177 | 94 | 183 | 104 | 232 | 16 | 696 |
| 16 | 99 | 168 | 261 | 94 | 220 | 20 | 674 |
| 17 | 61 | 150 | 330 | 47 | 181 | 36 | 619 |
| 18 | 84 | 81 | 196 | 80 | 184 | 55 | 544 |
| 19 | 118 | 92 | 309 | 71 | 173 | 81 | 671 |
| 20 | 126 | 92 | 310 | 61 | 193 | 101 | 690 |
| 21 | 143 | 81 | 297 | 95 | 184 | 123 | 719 |
| 22 | 88 | 76 | 317 | 88 | 169 | 149 | 662 |
| 23 | 65 | 38 | 210 | 93 | 183 | 172 | 551 |
| 24 | .. | 11 | 98 | 53 | 124 | 185 | 275 |
| 25 | 5 | 12 | 135 | 83 | 112 | 197 | 235 |
| 26 | 13 | 16 | 128 | 84 | 113 | 201 | 338 |
| 27 | 13 | 5 | 91 | 75 | 129 | 250 | 308 |
| 28 | 48 | 3 | 74 | 49 | 118 | 325 | 289 |
| 29 | .. | .. | 39 | 54 | 59 | 428 | 152 |
| 30 | 4 | .. | 36 | 48 | 53 | 507 | 141 |
| 31 | .. | .. | .. | .. | (42) | 532 | 125 |
| 32 | .. | .. | 18 | 60 | 30 | 588 | 108 |

The actual oviposition per female-day is:

| Age of culture in days | Average eggs per female-day | Average females per bottle |
|---|---|---|
| 1–5 | 38.6 | 2.0 |
| 6–10 | 24.1 | 2.0 |
| 11–15 | 3.24 | 73.0 |
| 16–20 | 0.83 | 118.8 |
| 21–25 | 0.55 | 102.2 |
| 26–30 | 0.21 | 65.6 |
| 31–35 | 0.00 | 18.2 |

Comparing the results, fertility is as follows:

|  | Eggs per female | Eggs per female-day |
|---|---|---|
| Actually observed . . . . . | 5.72 | 0.975 |
| If all eggs develop to maturity | 1.73 | 0.198 |
| If the initial conditions continue | 83.53 | 20.268 |

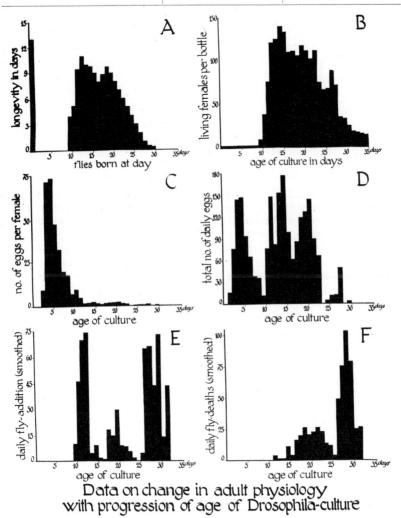

Data on change in adult physiology with progression of age of Drosophila-culture

Fig. 16. Changes of adult physiology with the progresssion of age in the *Drosophila* culture: A. Adult longevity. B. Number of breeding females per bottle. C. Eggs per female-day. D. Total eggs laid daily. E. Daily addition of adult flies. F. Daily deaths of adult flies.

The change of the age structure in the *Drosophila* bottle is illustrated in Fig. 4, B.

The growing population is almost pyramid-shaped, the stable population bell-shaped, and the contracting population urn-shaped in age structure. This is the structure known from human populations as well as from bee populations. In our present analysis the age-structure is of special interest, because it confirms our opinion that the logistic curve of population growth within the *Drosophila* bottle is really a superposition of two growth-periods. On the 10th day the age-structure shows the first beginning of a dwindling population, whereas on the 15th and 20th days the pyramid shape is again restored, after the oviposition of the adults of the first filial generation has started. On the 25th day dwindling has started again, and now leads to the final contraction, the progression of which is well illustrated by the age-structure on the 30th and still more so on the 35th day.

When ALLEE [4a] claimed that the *Drosophila* chapter of our "Problems" had been prematurely incorporated into a book, he was a sound prophet. Among the many improvements by far the most important is doubtless that of ROBERTSON & SANG [5]. They probed the implicit assumption of all earlier workers that the food supply for all stages of development was sufficient throughout the duration of the experiment, at least as long as the stable maximum population was maintained. They actually found that the yeast in the agar layer at the bottom of the bottle, the sole nourishment of the maggots and the main food of the adult flies, disappears rapidly when many maggots feed on the agar, and that the yeast is unable to recover its previous standard because its place is taken by a pink *Oidium*-fungus, unsuited to serve as larval food. It is only under such conditions of starvation that the population-density affects the rate of egg-laying as described before. There is no such effect when sufficient yeast is added daily to the agar.

**Table XII.**
Variation in the number of eggs laid by *Drosophila* during the first ten days according to food-supply conditions and the number of pairs in the bottle.

| Food-supply conditions | No. of *Drosophila*-pairs in bottle | | | | | | |
| --- | --- | --- | --- | --- | --- | --- | --- |
| | 1 | 2 | 4 | 8 | 16 | 32 | 64 |
| Old-type experiment | 45 | 28 | 31 | 26 | 6 | 5 | 4 |
| No yeast at all in bottle | 53 | 26 | 8 | 12 | 18 | 16 | 6 |
| Great surplus of yeast throughout experiment | 102 | 121 | 148 | 204 | 158 | – | 100 |

These are the results of our own experiments with which we achieved full agreement with those of ROBERTSON & SANG [6]. Our experiments on *Drosophila* encouraged us to study the influence of larval density and of food density (yeast) on the development of *Aedes aegypti*[7]. Food density below 0.6 g of yeast per unit increased the development considerably, while food in abundance had no such influence. The second part of the table demonstrates that varying larval densities of the corresponding same food density had no influence at all.

**Table XIII.**

Influence of food- and larval density on the growth of *Aedes aegypti*

| Temperature | g. of yeast per unit | no. of larvae | Average development in days |
|---|---|---|---|
| 24° C | 0.1 | 30 | 22 |
|  | 0.2 | 30 | 20 |
|  | 0.6 | 30 | 15 |
|  | 1.0 | 30 | 15 |
|  | 1.4 | 30 | 15 |
| 28° C | 0.2 | 30 | 10 |
|  | 0.4 | 60 | 11 |
|  | 0.6 | 90 | 11 |
|  | 0.8 | 120 | 11 |
|  | 1.0 | 150 | 12 |
|  | 1.2 | 180 | 11 |

These experiments again prove that the limiting factor determining the speed of development is not the population-density but the food quantity.

These and other analyses, by CHIANG & HUDSON[8], have stimulated intensified research into other environmental factors in the apparently simple environment of the *Drosophila* bottle, such as chemical changes and the acidity or otherwise of the agar. In addition, an increasing number of intrinsic and demographic factors have been found to complicate their interplay still more. Some of them are reviewed[8]:

"The reduction in population increase takes place in the cultures before any change in adult density. It is only by such factors as the reduction of fecundity and egg viability due to larval activity, the larval mortality due to intra-stage interference and food shortage, and in addition, the effect of adult density in reducing fecundity, that the population growth rate can be brought down to the level shown by the census figures. And since there is an increase in number after the majority of the larvae have pupated, the qualitative change in yeast is not likely to be the important cause of a drop in production. These findings based on experimental methods supplement Pearl's explanation and demonstrate some of Bodenheimer's postulations regarding the dynamics of *Drosophila*-population in growth.

"Although the growth of *Drosophila*-populations in a culture bottle has the same sigmoid curve as the growth of human populations and that of yeast, the dynamics involved are entirely different. With the magnitude of reproductive capacity and of food consumption exhibited by *Drosophila*, a much larger environment, or a much more highly replaceable food resource than that available in the milk bottle is necessary, if the population is to follow the same type of sigmoid curve as that for human or yeast populations. Pearl's (upper) asymptote is a temporary one, and the sigmoid curve obtained is only the portrait of a prolific family growth rather than population grcwth in the usual sense. The *Drosophila*-population growth in the culture bottle is actually more demonstrative of the outbreak-crash type which is also very common for population growth in nature. At the initial stage, the reproductive potential is almost fully expressed and its result is an outbreak. The increase in numbers takes place much before any density-dependent factors are effective, so that the production soon overshoots the carrying capacity of the environment and a crash follows. Under these conditions the establishment of population equilibria is nearly impossible."

Despite these last remarks of Chiang & Hudson, the logistic curve remains a fitting description of the actual population growth of *Drosophila* and of other animals, including man. But the importance of the biological density factor as its regulating mechanism is now much reduced. Even in the very simple and relatively stable case of the *Drosophila*-bottle the extrinsic and intrinsic conditions are extremely complex. The density factor retains its high primary importance only under conditions of starvation, when larval survival would be insignificant in any case, or in a change of behaviour as group effect.

There remains the question of what brings about the identical curve in all the manifold types of population growth. There is apparently only one answer. This well-defined mathematical formula fits for all cases of growing populations only if all factors stimulating population-growth, as well as all those inhibiting it, act – in spite of all their intrinsic and extrinsic diversities – as though only one combined stimulating factor were acting against one inhibiting factor. Then, and then only, is the logistic curve a necessary outcome. The fact that every growth is adequately described by a logistic curve is adequate proof that this is actually the case. There can therefore be no doubt that the logistic curve of growth is a purely formalistic rule. Not much remains of the regulating density-factor which we all accepted thirty years ago as the explanation of the logistic curve of population growth.

This leads to a significant and healthy conclusion: the total product of the inhibiting as well as of the stimulating factors must be dissected into their individual components to achieve an approximate evaluation of each. The similarity of two logistic curves does not justify a conclusion as to the identity of factors determining each curve. Identical curves will, as a rule, be produced under very different conditions and by widely differing factors.

Thus, in the case of human demography: every individual growth curve of a group or of a country must be analysed on its own merits. Some decades ago many human demographers, among them the author, tried to interpret various demographic, cultural and economic phenomena of history as consequences of the position of the population concerned on its growth curve, i.e. as depending upon a "density-factor": liberal tendencies in economic and spiritual attitudes were correlated with the middle third of the curve, and totalitarian ones with the upper third. Rapid deterioration of the environment was supposed to lead to a rapid jump from liberal to totalitarian economy and attitude when the vital space of the individual had suddenly become restricted below the threshold where free movement and initiative is possible and useful for the community. But even the decrease of reproduction in huge capitals – heretofore considered primarily the result of a biological "density-factor" – calls for interpretation anew, and an individual analysis in every case. The high birthrate in the wealthy strata in capitals of primitive societies such as Bagdad, the low birthrate of workers in present-day industrial centres, the recent rapid rise of the birthrate in academic circles in many parts of Europe and in all classes of the Parisian population – all these and many other phenomena must be analysed with regard to their individual economic, psychological, etc. factors. They are not merely expression of a – let us confess it now – largely mystical "biological density factor". Many human demographers have in any case never accepted this density factor as a primary regulator, although it seemed so self-evident from a biological point of view[9].

Comparison and Conclusions. This analysis of growth of *Drosophila* populations may be compared with the population of two other insects growing in a limited space.

We shall begin with CHAPMAN's classical experiments on the flour beetle *Tribolium confusum* and the able analysis and variation of these experiments by STANLEY, HOLDAWAY, PARK and others[10]. Table XIV gives the comparative analysis for this beetle bred at 32°C and 75% R.H. The constants are as follows: egg 4.42 days; larva 17.34 days; pupa 5.38 days; eggs per female-day 10.73; sex-ratio about equal.

The analysis shows that within the period of observation 298,679 eggs should have been laid. 22,631 eggs, i.e. 7.6% of the potential

**Table XIV.**

Population trend and analysis of *Tribolium confusum* at 32° C

| Day | A<br>Total<br>Population | B<br>Adult<br>Population | C<br>Potential<br>increase of<br>population | D<br>Actual no. of<br>eggs per day |
|---|---|---|---|---|
| 0 | 16 | 16 | .. | .. |
| 10 | 568 | 15 | 858 | 802 |
| 20 | 563 | 14 | 751 | 237 |
| 30 | 596 | 92 | 2,468 | 168 |
| 40 | 760 | 350 | 9,389 | 127 |
| 50 | 791 | 362 | 7,769 | 344 |
| 60 | 844 | 373 | 20,065 | 1,040 |
| 70 | 1,056 | 362 | 19,421 | 1,559 |
| 80 | 1,138 | 364 | 19,529 | 1,722 |
| 90 | 1,286 | 361 | 19,421 | 2,068 |
| 100 | 1,299 | 357 | 19,207 | 2,106 |
| 110 | 1,498 | 350 | 18,777 | 2,515 |
| 120 | 1,095 | 350 | 18,777 | 1,471 |
| 130 | 1,013 | 343 | 18,455 | 1,340 |
| 140 | 1,213 | 325 | 17,489 | 1,887 |
| 150 | 1,137 | 319 | 17,168 | 1,652 |
| 160 | 853 | 297 | 15,988 | 1,053 |
| 170 | 883 | 272 | 14,593 | 1,254 |
| 180 | 1,433 | 266 | 14,271 | 2,204 |
| 190 | 836 | 241 | 12,983 | 796 |
| 200 | 763 | 209 | 11,367 | 782 |

egg-production were counted. From these 22,631 eggs 1,079 larvae hatched, of which 601 pupated. Between 386 to 600 adults hatched from these eggs, probably about 400.

Following the observation that larvae and adults of *Tribolium* willingly eat their own eggs (but do not search for such food!), CHAPMAN[10] and his co-workers concluded that the increased chance of meeting eggs leads to a constant heavy decimation which is the real reason for the stable adult population often observed for quite a considerable period in these experiments. We may assume that the difference between eggs actually laid and the number of larvae observed is due largely to feeding of the adults on the eggs. But even if, in addition, the same amount or twice as much is destroyed in this way, only a fraction of the potential egg-production is accounted for. This has been demonstrated also by MACLAGAN[12] from CHAPMAN's own data:

| Beetles per g flour... | 2 | 4 | 8 | 16 | 32 | 64 |
|---|---|---|---|---|---|---|
| Eggs per beetle-day.. | 1.56 | 3.80 | 3.00 | 1.56 | 1.04 | 0.80 |

In *Tribolium* as well as in *Drosophila* there exists a clear correlation between population density and fertility. Below and above the optimal density fertility is reduced and the potential fertility is not realized owing to population pressure. This factor accounts no doubt for the largest part of the unrealized potential increase. It is of great interest to gather, again, from this analysis that environmental resistance as well as intensity of struggle for existence do not

**Table XV.**
Population trend and analysis in honey-bee populations in a single-walled bee hive at Kiryat Anavim, 1936[11].

| Date | Total population | Reproducing females | Potential increase of population | Actual no. of eggs laid per period |
|---|---|---|---|---|
| 1.I | 11,800 | 1 | 33,000 | 4,975 |
| 11.I | 16,150 | 1 | 30,000 | 4,780 |
| 21.I | 21,620 | 1 | 30,000 | 5,850 |
| 1.II | 27,430 | 1 | 33,000 | 7,065 |
| 11.II | 34,545 | 1 | 30,000 | 9,500 |
| 21.II | 41,015 | 1 | 30,000 | 9,500 |
| 1.III | 48,605 | 1 | 24,000 | 9,940 |
| 11.III | 53,935 | 1 | 30,000 | 11,990 |
| 21.III | 61,355 | 1 | 30,000 | 9,740 |
| 1.IV | 65,980 | 1 | 33,000 | 11,375 |
| 11.IV | 65,325 | 1 | 30,000 | 9,390 |
| 21.IV | 66,285 | 1 | 30,000 | 9,810 |
| 1.V | 63,760 | 1 | 30,000 | 9,130 |
| 11.V | 61,015 | 1 | 30,000 | 9,335 |
| 21.V | 60,370 | 1 | 30,000 | 9,275 |
| 1.VI | 57,915 | 1 | 33,000 | 9,320 |
| 11.VI | 54,015 | 1 | 30,000 | 4,605 |
| 21.VI | 46,765 | 1 | 30,000 | 4,030 |
| 1.VII | 42,895 | 1 | 30,000 | 5,075 |
| 11.VII | 29,230 | 1 | 30,000 | 5,570 |
| 21.VII | 31,815 | 1 | 30,000 | 5,685 |
| 1.VIII | 32,660 | 1 | 33,000 | 7,570 |
| 11.VIII | 35,135 | 1 | 30,000 | 6,355 |
| 21.VIII | 36,990 | 1 | 30,000 | 6,110 |
| 1.IX | 37,855 | 1 | 33,000 | 5,925 |
| 11.IX | 37,690 | 1 | 30,000 | 5,170 |
| 21.IX | 35,005 | 1 | 30,000 | 4,575 |
| 1.X | 32,875 | 1 | 30,000 | 4,365 |
| 11.X | 31,175 | 1 | 30,000 | 3,990 |
| 21.X | 27,665 | 1 | 30,000 | 3,195 |
| 1.XI | 23,070 | 1 | 33,000 | 0 |
| 11.XI | 17,270 | 1 | 30,000 | 1,100 |
| 21.XI | 14,560 | 1 | 30,000 | 1,320 |
| 1.XII | 11,895 | 1 | 30,000 | 1,630 |
| 11.XII | 10,005 | 1 | 30,000 | 1,270 |
| 21.XII | 7,345 | 1 | 30,000 | 1,630 |

increase indefinitely, but remain fluctuating at values below saturation. After the population peak is passed, both factors tend to relax, as a rule.

Analysis of research on *Tribolium* shows again how careful one should be in applying a simplified mathematical analysis to rather complex biological phenomena. STANLEY treated the given facts successfully as an indefinite periodic fluctuation between predator and prey, applying formulae of chance meeting of gas molecules. The actual analysis shows that this conception is rather incomplete and fails to explain many important numerical features. But it suggests that the biotic factors of population control, to which belong the population density as well as the food relation, are interchangeable in their effect and may possibly be treated successfully by mathematical analysis as one compound factor.

The analysis of the population trend of the honey-bee reveals another type of population growth. Reproduction is here independent of the number of adults produced, as one reproducing female only is to be found within each hive. Extrinsic conditions, as abundance of nectar and pollen, climate and empty space for brood-rearing, act as limitations of the potential increase[11]. As 3,000 eggs laid was the observed maximum for one oviposition day, this has been regarded as the potential increase. Table XV contains the data and their analysis.

Far from having explained the dynamics of the population trend within the *Drosophila* bottle, the complete population count has raised many new problems and shows the facts from another angle. The experimentation is only at its beginning, but a few facts of outstanding importance may be pointed out:

(1) The number of adults hatching is only a fraction of the eggs laid.

(2) A prolongation of development is to be expected in the later stages of the culture.

(3) Heavy adult hatching occurs just at the final stage, followed by death even before oviposition begins.

(4) The final contraction is due to intoxication.

(5) Age structure shifts in the same way as that in changing populations of man or honey-bee.

(6) Intensity of struggle for existence, as conceived by us, does not grow indefinitely. After development has passed the population peak it decreases or remains constant. An increase in environmental resistance indicates that other factors, besides population pressure, participate in the resistance.

## 2. Aphid populations

Aphids provide a boundless source of information concerning population ecology. At one end, we find holocyclic species which are more or less monophagous, non-migrating (autoecous) species. The favourable season of climate and host, organically correlated, is utilized by very few parthenogenetic generations. The long unfavourable season of climate and host leads over, after one or a few generations, to the gamic cycle and the diapause eggs. Gamic morphes become conspicuously abundant only for a very short season, if at all. The number of eggs laid largely determines the size of the population the following year. This number of eggs often depends very much on the weather at the oviposition period. Storms, heavy rains, or any adverse weather may greatly reduce this number[13].

At the other end, we find polyphagous species with only thelytokous (parthenogenetic) reproduction, which tend to utilize intensively and speedily by rapid increase every favourable food source and climate condition. For many successive weeks and even months we may find them in astonishing abundance, mainly depending upon the abundance of hosts in suitable condition. Unfavourable climate and host condition (shorter days, a diminishing trend of nightly minima of temperature, high temperatures, etc.) favour directly or indirectly the production of viviparae alatae or of sexuparae. In either case (by active or by passive flight, or by production of diapause eggs or hibernating larvae) this represents an effort to avoid the unfavourable environment.

As soon as aphids become abundant they form one of nature's staple foods. As Brehm says of the hare: "Alles, alles will ihn fressen." Many species of ladybeetles, lacewings, syrphids and other flies, mites and predators, as well as a great number of parasitic wasps, join in their destruction. Yet the reproductive potential of the aphids in their favourable season is so enormous, that even their efficient local enemies very often cannot control them. What makes them so abundant and conspicuous is another fact: the peak of the number of these enemies almost always lags weeks behind that of their prey. This means that enemy action almost invariably becomes important and significant only when the population peak of the aphids has for weeks been on its downward trend. Then we observe everywhere an apparently enormous decimation by predators and parasites, but only in the already speedily shrinking aphid populations. We were able to demonstrate clearly that in certain cases the physiological changes in the hosts are the primary factor leading to the sudden decline. Where enemies almost lead to extermination of local aphid colonies, the species is maintained by the rapid establishment of new colonies. Whenever aphids are rare, or hidden, the enemy component is negligible.

Between these extreme fluctuations of moderate and abundant aphid populations, we find such an embarrassing variety of intermediate types of life cycles, of different life cycles of the same species in different regions and even years, that the categories constructed by the human mind cannot do justice to the enormous variety of adaptations. The study of aphid populations will satisfy our quest for many essential and ever present principles, such as, for example, an explanation for the fact that dense populations cannot persist without passing into a very definite population ebb. Either physiological changes of host conditions occur; or climatic changes such as heavy rains, storms, cold or heat; or enemies appear in numbers; or diseases. While unfavourable climate is often one of the most important and decisive lethal factors, it is not a controlling factor, as its effect is entirely independent of host density. Host condition and enemies are often controlling factors, but in aphids the former predominates, so that when the enemy populations have grown to significant size the aphid populations are already in steep decline due to unfavourable host changes.

The basic conclusion to be drawn is that there is no easy and dogmatic solution for any problem; that the phenomena witnessed need the support of many systematic observations for full understanding and analysis; that survival as the test of adaptation has been attained in aphids in a greater variety of ways than is known in any other group of animals.

In very mobile and dispersed aphids, such as *Eulachnus tuberculostemmata* on Tamarisk, enemies have practically no influence at all on the populations. In the abundant *Myzus persicae* the young larvae disperse from the very moment of their birth, while in the more sessile *Cinara palaestinae, Brevicoryne brassicae* or *Aphis nerii* colonies aggregate and sooner or later offer excellent opportunities for extermination by enemies. In the meantime, however, flying thelytokous (parthenogenetic) females, produced often by the group-effect, have already established many new colonies, thus maintaining the continuity of the species. *Myzus persicae*, the polyphagous and mobile gypsy aphid, has a tendency to remain in the profitable thelytokous (parthenogenetic) condition as a vagrant throughout the year, wherever it can.

Only a very small fraction of the populations shows a facultative amphigonic cycle[14]. In December 1950 some sexuales and eggs on apple trees in the coastal area were observed for the first time. The colonies founded by these autumn remigrantes were reinforced by viviparae alatae in March-April arriving from many herbs. By May the apple trees were abandoned and remained free of aphids until the following autumn. The apple tree is the only rosaceous fruit tree which does not shed its leaves completely in early winter, and which offers a suitable food to the sexuales. Mirobolan plums which gave young leaves in winter were infested by viviparae alatae, and peach, pear, apricot and almond leaves were invaded in early spring by viviparae alatae. None of the viviparae feed on old leaves, but gynoparae and sexuales do. The *Myzus* populations on trees are never large in Israel. Apparently the main population peak in most secondary hosts (e.g. potato) is in spring, with a minor peak in late autumn. WILLCOCKS' observation (1937) in Egypt of apterous gamic females and winged males as well as eggs on *Prunus* in February and March is interesting. This is an interesting adaptation of this well-known European peach aphid to a new gamic host induced by local conditions.

And now a few notes on some of these parasites. *Aphelinus mali* HALD. is a North American parasite of *Eriosoma*, which has been by now introduced inadvertently or for biological control into almost all countries where *Eriosoma lanigerum* appears as a pest. It was introduced into Egypt after 1930, we obtained it from there in 1935, and sent it on further to Turkey and Syria[15].

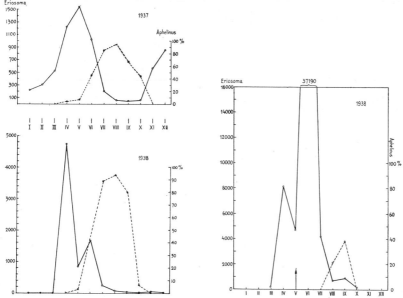

Fig. 17. Population trend of the Woolly Aphis *Eriosoma lanigerum* (full line) and the seasonal trend of its parasitation (in percentages) by the wasp *Aphelinus mali* (interrupted line). Above at Kiryat Anavim (west of Jerusalem) where the parasite was introduced since some years; below at Atarot (north of Jerusalem) where a small introduction was made in May 1938. In both localities did the aphid population strongly contract before any influence of parasitation could have been the cause (From BODENHEIMER, 1947).

In Kiryat Anavim diapause begins in September, due to the declining trend of temperature minima, while the effective temperatures would still permit the development of two further generations. The hibernating wasps hatch in March. Their development in summer lasts about 12 days. The wasps oviposit preferably in the young larvae on the periphery of the host's colonies. Drops of body liquid extruding from the puncture hole are eagerly lapped up by the ovipositing females. Even in case of multiple oviposition into one aphid, only one wasp hatches. The infested aphids become inflated and shining black.

The threshold of development is at 8.6° C, i.e. about 4.4° C higher than that of the host (4.2° C). Thus *Aphelinus* reaches abundance only in June when the *Eriosoma* population is already in its steep decline. Figure 17 shows the seasonal incidence of *Eriosoma* and *Aphelinus* populations at Kiryat Anavim in 1937 and 1938. It seems doubtful, even from a study of these curves, whether *Aphelinus* had any significant influence on the downward trend of *Eriosoma*. Yet the relatively large number of parasitized aphids just at the time when the *Eriosoma* grow rare is always a "convincing" argument to the layman for the very active part the parasites played in the sudden heavy decimation of abundant aphid populations.

We have, fortunately, an interesting check in some counts at Atarot in 1938, where we did not wish to introduce *Aphelinus*. Yet one settler infested his garden in June with *Aphelinus*. Figure 17 shows clearly that the late and little *Aphelinus* parasitization could not possibly have brought about the steep downward trend of *Eriosoma* in early summer.

This seasonal behaviour of *Aphelinus* confirms again the decisive lag observed in all aphid parasites. During the population ebb of the host parasitism is very low, which in this case happens in winter diapause. *Aphelinus mali* was apparently able to establish itself in all areas of the middle East where it was introduced, without, however, fulfilling in any place the hopes set upon it for biological control.

These few remarks should suffice to whet the appetite for further knowledge of the extremely versatile and complex ecology of aphids.

### 3. The populations of the Desert Locust
*(Schistocerca gregaria)*, **the locust without a home**

As far back as Biblical times (Exodus, Joel, Nahum), and even before then, according to earlier Egyptian and Mesopotamian accounts, the recurrent invasions of the Desert Locust have been dreaded catastrophes for the agriculturist of the Middle East, North Africa, Northwestern India, etc., although it was at the same time "manna from Heaven" for the ever-hungry nomads of these areas. Nonetheless, it has only been within the past few years that the great patience of B.P. UVAROV (of the Anti-Locust Research Centre, London) is bearing fruit, giving us insight into the dynamics of the migration, the numerical fluctuation and the changes of behaviour of this locust under various biotic and abiotic conditions.

The author has been privileged to share in a full generation of trial and error concerning this locust, but the knowledge acquired led to the discard of many earlier ideas but not to the full establishment of new ones. When we began our studies there was only one good life-history, worked out by VOSSELER[16] in East Africa. It was then still assumed that the Desert Locust is at home all over its outbreak areas, even if it was extremely rare in the outbreak intervals.

B. P. UVAROV initiated a new period with the phase theory of the European Locust *(Locusta migratoria)*.[17] It was many years, however, till it was realized that phases, solitary and gregarious forms, occur with all locusts. It took still longer to ascertain that the morphological and colour phase is only the expression of important physiological and fertility changes, by no means coincident with the changes of morphology. Important physiological investigations[18] revealed the surprising fact that Desert Locusts oviposit in wet soil only and, even more surprising, that the freshly laid egg is extremely sensitive to humidity below saturation. The one-week old egg becomes progressively drought-resistant, and weather has very rarely a serious direct effect on the mortality of *Schistocerca* in the larval and adult stages.

We had assumed that the original and permanent home of *Schistocerca* was the short-grass savannah of the Northern Sudan right across Africa, the corresponding parts of Arabia and Northwestern India. All these regions have early summer rains. We further assumed that after rapid sexual maturation of the surviving adults, oviposition began soon after the onset of the rains, the larvae utilising the still fresh green of the vegetation, while the immature adults – decimated by birds, lizards, etc. – survived until the following rains, hiding in wadis, in dense vegetation or in mountain areas with more precipitation. We failed, however, to find them in all these areas.

The theory of *Schistocerca* outbreaks was then that in regional border areas – for example, near the Saharo-Sindian Red Sea Coast – early Winter rains penetrate farther inland and fall where normally only the Sudanian Summer rains would fall. This meant sexual stimulation of a much higher population of *Schistocerca* adults and propagation of a second generation. In many cases this led to high concentration, followed by gregarisation of the hoppers. The hatching adults were able to migrate from areas of summer rains to those of early winter rain; the outbreak was on its way and reached greater populations and greater migrations every year. All the invasion areas, wholly free of locusts in the intervals, were now successively invaded, and at the peak of the outbreak as many as three annual generations might develop[19].

Now we know that all these "permanent solitary populations" in the northern Sudan – if they exist at all – have no importance in connection with the great outbreaks and migrations, extending from the Mediterranean region over the Middle East, U.S.R.R., Central Asia and Northwestern Indian in the north, to Ethiopia, Somaliland and East Africa in the south. The Desert Locust dies out in these areas and very few individuals may be found there in the years immediately preceding or following a major outbreak.

We mentioned before that during outbreaks and after we distinguish two different phases of the Desert Locust, as in all other locusts. It seems that the morphological and colour changes are connected with higher activity, perhaps with higher basic metabolism. In *Schistocerca* the solitary phase is greenish in all its stages. The much more mobile gregarious phase has vivid black-yellow-pink larvae, with adults who change from an early brownish to a pinkish and, at sexual maturity, to a yellow colour. Morphological changes occur primarily in the greater length of the elytra in relation to the stable hind-femora. Most important are behaviour changes. The solitary locusts do not react to their fellows, are not attracted to them and show no tendency to gather into coherent groups. The gregarious phase possesses all these characteristics in a pronounced degree – it is much more active, often more fertile than the solitary female, and – depending on temperature and soil-humidity conditions – is able to produce two or three annual cycles of generations, in contrast to the one annual generation of the solitary locust.

Population density and phase transformation go hand in hand, crowding being required invariably to produce the full gregarious phase from the solitary one. We may distinguish the following three stages of behaviour development[20]:

(1) **Concentration** is the absolute increase of population density in a restricted area.
(2) **Aggregation** is the active grouping of individuals into bands in which fellow-individuals assume increasing importance in the sensory experience of every locust.
(3) **Gregarisation** is the more massive swarm formation, with speedy fusing of all meeting individuals, bands or swarms, all very active, inclined to wide-range migration, with definite production of the gregarious phase.

The local and casual increase and gregarisation of hopper bands in the wide area inhabited and invaded by *Schistocerca* are no longer considered of primary importance for gregarisation, swarm formation and swarm maintenance. The main accent of swarm formation is now on the adults.

The dynamics of weather help the formation of the flying swarms. In May 1950 a number of *Schistocerca* swarms was followed when it began to move from the Red Sea Coast of Saudi Arabia towards French Equatorial Africa, 2200 km away, where they arrived one month later. These and similar facts led R.C. RAINEY[21] to study certain seasonal migrations of winged Desert Locusts. Prolonged flights and the general importance of wind-direction for the locust flights have been recognised for some time. The whole of the migration across the Sudan in May occurred along the northeast monsoon. It became obvious that, whenever the wind speed is not less than 7. m. p.hr., every flight is with the wind, even those individuals or swarms which apparently head against it. A swarm which fails to maintain a definite direction, moving either in circles of in zigzags, will also move with the wind, even at lower wind intensities.

A synoptic chart, recording simultaneous weather observations over a wide area, reveals areas of air convergence. Convergence is an essential factor in the production of widespread heavy precipitation. One of the main areas of convergence affecting regions invaded by *Schistocerca* is the Intertropical Convergence Zone between the trade-wind and the monsoon currents, originating on opposite sides of the Equator, in Africa and India. The major displacements of locust swarms take place towards these areas of convergence. The subsequent concentration of the swarms and of the scattered locusts by such convergent winds can bring together most of the locusts from many thousands of square miles into a restricted zone, where the opposing winds meet (Fig. 18). Such concentrations then lead easily to aggregation and gregarisation. Even when such swarm formation occurs only every few years, it suffices to replace those swarms which have in the meantime disappeared and keep the cycle moving.

The late winter invasions into the Middle East are interruptions of the usual cold north-to-south winds by warmer spells which lead from south to north, bringing the locusts with them, as well as bands of *Pyrameis cardui*. The reliability of rainfall, leading to stability of outbreak areas, is very important for locust concentration. It is characteristic of *Schistocerca* that there is relative unstability of rains within the area of major migrations except for the marginal areas. Such concentrations – even when they occur only once in eleven or thirteen years – are now considered the prime motor of outbreaks, the already gregariously massed eggs with the hatching

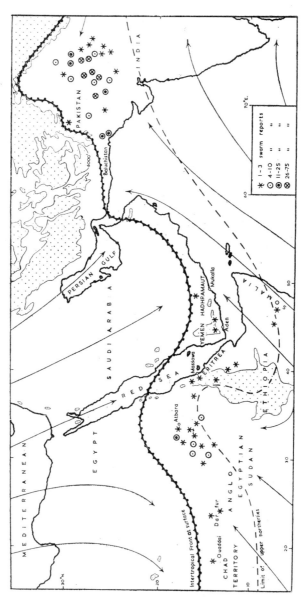

Fig. 18. Intertropical concentration zone of *Schistocerca gregaria* between the monsoon and the trade-winds. (From RAINEY, R. C., Weather and the Movements of Locust Swarms, a new Hypothesis. Nature, Lond., Vol. 168 (1951) page 1059).

of huge and concentrated hopper bands continuing the process of intensive gregarisation and outbreak.

There is no doubt that the continuance of a plague of locusts depends on the ability of swarms to utilize winds for their migrations and thus change the breeding areas of successive generations. Seasonal migrations of swarms have reached, in the case of the Desert Locust – according to UVAROV – an extent unknown to other locusts, enabling seasonal reproduction in widely distant areas, with different types of climate, but with similar weather at the time of breeding. Always on the move in search of suitable oviposition areas, the *Schistocerca* swarms move over vast areas, having no home, no generation ovipositing near its own hatching ground, and none of the following generations necessarily returning there.

We are thus led to abandon our previous theories – which still hold good for most other locusts – that all outbreaks begin their increase and aggregation in very definite areas, in the so-called gregarigenous centres, which are all on the borderlines of two different regions with other rain regimes and other types of vegetation. The ecotone, a zone of contact between two types of plant cover, accentuated often by human activities, is of increasing importance for the outbreaks of other locusts. *Schistocerca*, however, can dispense with the need for any fixed gregarigenous zone because of its almost incredible mobility and the enormous extent of normal migrations during outbreaks. We know that apart from conditions required for the laying and development of eggs and for the very young hoppers, the onset of a prolonged arid season is important as a primary motor for the beginning migrations of adult swarms. The amazingly great migratory vagility in quest for weather and vegetation favourable for maturation and development liberates the Desert Locust from practically any stability within the enormous ecologically available area. While the solitary phase is not more vagile than any ordinary grasshopper, we know today that long flights of the same adult swarm may proceed from the apparent centre in northeastern Sudan to Rio de Oro, or Morocco, to East Africa, or to India and Afghanistan in direct flight. Deprived of this tremendous vagility there could be no development of outbreaks, no long deplacements, no survival of the swarms in the invasion areas.

UVAROV[22] resumes its epidemiology as follows: "Even in the Sahara vast expanses of densely growing annual herbs at times offer ideal pasture for locusts and an increase in population (by immigration) is liable to occur. Such vegetation, however, is very short-lived and may begin to dry up even before the hoppers reach the adult stage, in which case the population would be concentrating in the greener patches in wadi beds and other depressions. In this way, hopper bands have been observed to form. Resulting adults, already with gregarious habits, have to leave the habitat where by that time there is neither food nor shelter and to migrate for breeding in another rainfall area. Their migration may be successful, or not; in the former case, a further population increase and greater degree of gregarisation would lead to swarm formation. Such conception of outbreak dynamics, involving a chain of seasonal breeding areas, most of them extremely uncertain, makes it evident that the whole process is subject to chance, and that it would not be reasonable to expect outbreaks to begin in any permanently localized area. Instead, it is necessary to envisage a series of events, combinations of which would make it possible for a nomadic population to increase and become concentrated in the course of several successive generations produced in different areas. Such possibilities are very many, but not unlimited, and the knowledge that seasonal movements and breeding of locusts are closely dependent on the weather dynamics provides a hope that in any given season the probable trends of locust population movements and the chance of its breeding can be inferred from synoptic analysis of weather in the critical areas and periods. Obviously, a close link between locust ecologists and meteorologists is as essential in the study of the Desert Locust outbreaks as is the analysis of its swarm migrations."

We have only a few minor remarks to make on this brilliant resumé: The outcome does not so much depend on chance as on the almost incredible migratory power of the adults. This reduces the chances enormously, as the probability of finding a suitable area for the next stage is very great indeed, in principle. The stiff-necked perseverence of the migrating adults makes this so. Too much importance is still given to the auxiliary aggregations of solitary hoppers. Last but not least, the behaviour of the Desert Locust, which we have touched upon in this analysis, brings it almost to perfection in an environment unfavourable for any stable animal. We conclude with the statement that – aside from a few generalities – everything said here concerns *Schistocerca* only, not any other locust, as every locust has its own environment, its own physiology, its own behaviour, and its own principles of outbreaks and crises, many very different indeed from those valid for the Desert Locust.*

## 4. The Epidemiology of Malaria

The epidemiology of human malaria is a model of the interrelation between the density of three or four groups of animal populations as well as of their dependence on factors such as climate, food, etc.[23].

The three types of human malaria are caused by three species of

---

* see note b page 267

Protozoans, called *Plasmodium vivax*, *Plasmodium malariae* and *Plasmodium falciferum*, belonging to the spore-forming group of that order. The *Plasmodium* develop in the human blood, where they form spores in the red corpuscles. Each successive cycle of development decomposes the red blood corpuscles within which they develop, at intervals of two or three days (or at irregular intervals), accompanied by attacks of high fever of short duration. The reproduction by spores is finally succeeded by the production of sexual cells – the gametes – of *Plasmodium*. These sexual forms are eventually sucked up with the blood of diseased human beings by blood-sucking insects, but only a few of the insects which suck human blood are able to offer conditions for the further development of the *Plasmodium*. The latter are destroyed and digested by all other insects except for a few species of *Anopheles* mosquitoes, where the *Plasmodium* pierce the wall of the mid-intestine, developing into new forms ready to infect new human beings. These then migrate into the proboscis of the mosquito.

This development takes time, as only after full maturation of the *Plasmodium* of this cycle within the female *Anopheles* (the males do not suck blood) can the latter infect anew, by sucking human blood. If the mosquito sucks cattle or other animals instead, as if often the case, the infestation form is destroyed within their blood without producing any symptoms of disease. It is only in human blood – the host suited to the development of the second part of the cycle – that they are able to develop and begin a new cycle in the red blood corpuscles of the new human host.

Cattle, with their heavy sweat, are much more attractive to mosquitoes than man. This means that dense agglomerations of domestic cattle close to human habitation detracts most of the danger from malaria without endangering the animals. A sudden heavy reduction of cattle population, in times of drought and famine consequently increases the danger of a local malaria outbreak enormously, because of the greater concentration of mosquitoes upon man. Furthermore, the normal resistance of well-fed men is drowned concurrently among the hungry human population – the cause of the catastrophic malaria outbreak in the Volga area in 1918, and of many outbreaks in the Punjab.

Human malaria thus involves a rather complicated ecological chain and depends, among others, upon the relative density of the *Plasmodium* (the malarial agent), the *Anopheles* mosquito (the carriers of the agent), and the receptors. The latter are sensitive to infection (in the case of man) or insensitive (in the case of cattle and other animals).

Climatic and other environmental conditions influence the intensity of the *Plasmodium* and *Anopheles* factors greatly. Bodies of water are required for the development of the larvae. Each of the dangerous malaria carriers has its special requirements concerning the type of water in which it breeds, as well as varying lower temperature threshold of development, ranging between 8° C to 17° C in the species so far studied. As a result, the continuous development of a number of generations of mosquitoes can take place only in warm countries or during the warm season of cooler regions. No development of any *Plasmodium* can take place below a temperature of 17° C. At lower temperatures the *Plasmodium* ingested by the mosquitoes are digested and destroyed. The Anopheline cycle of *Plasmodium* can develop only at temperatures above 17° C, which means that only at higher temperatures can new mosquitoes actually be newly infested. The quantity of new infectious *Plasmodium* developing within the mosquitoes also depends on the age of the female *Anopheles* at the time they suck the infected man (the malaria carrier). In the mosquitoes which die less than seven to ten days after the act of sucking – this is the period usually needed for the development of the Anopheline cycle of *Plasmodium* to gain maturity – no new infestation of man can take place. These doubtless represent a very high percentage of all infested mosquitoes.

Even if all the above-mentioned conditions exist, the onset of a malaria epidemic is by no means a foregone conclusion. A moderately dense population of suitable mosquitoes can be relatively unimportant for human malaria if a large cattle population in the neighbourhood of man attracts the overwhelming majority of the mosquitoes. The reverse is also true. This means that all factors, animate as well as inanimate, must be above a certain threshold – considerably higher than the minimum threshold for survival – before an epidemic can develop.

In Northern Europe, for example, a high density of the proper mosquitoes does not produce any malaria epidemic nowadays because, among other reasons

(1) the cool summer does not permit the development of the Anopheline cycle of *Plasmodium* every year;
(2) the intensive cultivation of the soil has led to a considerable lowering of the groundwater level, thereby reducing greatly the number of suitable breeding grounds for the mosquitoes;
(3) the number of human malaria carriers is usually very small.

Great changes in any of the factors involved may, of course, radically alter this favourable situation of the region concerned.

In recent years decisive steps have been taken in countries favouring malaria outbreaks. These include the efficient organisation of malaria control by killing the *Plasmodium* by proper treatment such as the supervised distribution of quinine and similar preparations, and the rigorous destruction of larvae and adult mosquitoes in human settlements by the use of D.D.T., petrolization, etc.

These measures made it possible to conduct warfare during the Second World War in highly dangerous malaria areas without any great loss by malaria among the armies. In certain countries, like

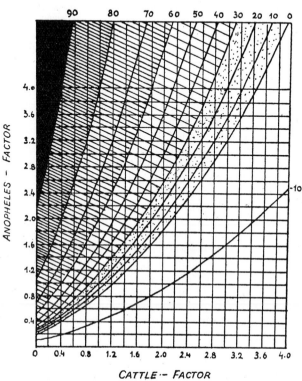

Fig. 19. The dependency of malaria-epidemiology on the intensity of the number of *Anopheles* present (*Anopheles* factor) and the intensity of cattle detracting these *Anopheles* from man (cattle factor). The histogram intends to show that below a certain effect of these or other factors, single infections are possible, but no malaria outbreak (From MARTINI).

Israel, malaria as an epidemic has been practically eradicated, but it remains a potential menace just as soon as preventive measures are interrupted. The mosquitoes may, of course, become immune to D.D.T., but other effective types of insecticides are being developed for such eventualities.

The famous major source of Roman malaria is the Campagna swamp area. According to CELLI, the fluctuations of malaria in Rome paralleled throughout its history the secular trends of precipitations correlated with the groundwater level (high at periods of heavy rain, and vice versa)[24].

The breeding habits of the various *Anopheles* species in various regions are of great importance for the local malaria situation, as they determine the number of *Anopheles* in close contact with man. In the city of Jerusalem a very serious malaria was endemic, with a heavy annual mortality, produced by *Anopheles bifurcatus*[25]. In Europe this mosquito breeds in forests and rarely comes into contact with man – thus it has no importance for human malaria. In hot dry Jerusalem environmental conditions forced this European forest mosquito to breed in the cool cisterns, thus remaining in very close contact to man. Since the inauguration of regular petrolization of all cisterns, in 1919, Jerusalem freed itself of malaria. Beisan, in the north of Israel, was also known for its serious malaria. This small town is surrounded on all sides by extensive swamps and it was therefore assumed that to rid it of malaria would be extremely expensive. BUXTON[26], however, showed that the *Anopheles* species inhabiting these swamps were not of importance in connection with the local malaria. One species only – living and developing in some of the rivulets running through the town – was responsible. The sanitation of these streams proved inexpensive and was an immediate success.

These are only a very few details of the extended and complicated epidemiology of human malaria, which serve as a good illustration of the importance of ecological research for health, agriculture, human demography, fisheries, as well as for pure science.

It is known that certain environmental conditions do not permit the development of a malaria outbreak even when all other factors are at the optimum stage. At temperatures below $16°$ C the *Plasmodium* factor is missing; when the suitable *Anopheles* is absent in the proper season, the carrier is missing; and in the absence of human *Plasmodium*-carriers, the human factor is lacking. Therefore, if one decisive link of the malaria chain is missing, no malaria epidemic can develop and the *Plasmodium* will die out.

Frequently, even if all factors are above the threshold, conditions do not permit the development of an outbreak although new sporadic malaria cases may appear. In Fig. 19 the *Anopheles*-factor refers to the number of mosquitoes present[27]. The cattle factor indicates the greater attraction towards cattle for the mosquitoes than towards man. The higher the *Anopheles* factor, the smaller the cattle factor, the more favourable are conditions for a malaria outbreak. In the entire unshaded field of Fig. 19 no malaria epidemic will develop. Malaria may develop only after the passing of a certain threshold by the rise in the number of *Anopheles* or the lowering of the number of cattle, and then only if a further change in the same direction takes place will there be conditions for a malaria epidemic.

Fig. 19 also demonstrates that an increase in the number of *Anopheles* is of little consequence close to the lower asymptote.

Nonetheless, after passing the threshold every slight increase of that factor changes the malaria situation considerably. The same is true of the cattle factor and of the degree of *Plasmodium* infestation of the human population. A medium initial *Plasmodium* infestation induces the quickest malaria infestation, which increase remains much lower at an initially very low or very high infestation.

R. Ross[28] has made a number of interesting calculations regarding malaria epidemiology:

(1) The presence of human *Plasmodium* carriers and of *Anopheles* beneath the threshold cannot produce an outbreak. If, for example, there is one *Plasmodium* carrier among 1000 men, and the average of all *Anopheles* is 48 per person, one quarter = 12 of these *Anopheles* may suck on the *Plasmodium* carrier; one third = 4 of these 12 mosquitoes live long enough to mature the *Plasmodia* within the *Anopheles* and one quarter of these (= 1) may infest successfully a healthy human being. Thus, of the 48,000 *Anopheles* present, only one infests a new man.

(2) Even if the initial infestation is high, the chances for a new infestation are much smaller than one would assume. If 500 among 1000 men have malaria and the malaria situation remains stable at 80 *Anopheles* per man, then a change in the number of mosquitoes will have the following consequences:

| Number of Anopheles per man | Months after the change: | | | | | Result |
|---|---|---|---|---|---|---|
| | 0 | 1 | 2 | 3 | 4 | |
| 100 | 500 | 525 | 544 | 560 | 571 | 600 malaria cases |
| 60 | 500 | 475 | 455 | 433 | 424 | 330 malaria cases |
| 40 | 500 | 450 | 409 | 376 | 348 | 0 malaria cases |

Where the factors remain stable, the percentage of new malaria infestation would be constant, a figure between 0 and 100%, and the latter, a 100% infestation, is a very exceptional result.

Each year and locality have their own malaria situation. The main components of infectiosity are *Anopheles* and cattle factor, resulting in the formula

$$a = q(1 + v)^{-2} \cdot R.$$

The symbols are: a: infectiosity; q: percentage of population which becomes free of *Plasmodium* by medical treatment; v: cattle factor, showing the great intensity of cattle in comparison to that of human attraction (1), and enters the formula in the square; R: all other malariogenous factors except man.

We conclude with a table showing the changes when in a country hitherto low in malaria the malariogenous factors (a) increase, and consequently a new equilibrium is established:

| Curve | Infectiosity (a) | Expectation of complete convalescence (q) | New equilibrium: % of malaria |
|---|---|---|---|
| 1 | 2 | 0.2 | 90% |
| 2 | 0.33 | 0.5 | 10% |
| 3 | 10 | 0.2 | 98% |
| 4 | 4 | 0.4 | 90% |

Case (2) illustrates how when a general infestation is reduced from 100% to 10%, a is reduced to one third, and q raised to 0.5.

## 5. Vole Population in Israel

Voles *(Microtus guentheri)* have been known since Biblical times (see Samuel I, 4), and ARISTOTLE gave a classical description of vole outbreaks in Greece in his *History of Animals* (580b). It is this very same Levant vole which LÖFFLER revealed in Thessaly as the bearer of an epidemic he called mice-typhoid[30]. Similar voles of the same genus are to be found all over Europe, North Asia and North America. They have forced our attention throughout the ages because of their cyclic increases, of primary importance to agriculture and economy. In the days of the Crusaders (1202/04) one such outbreak, combined with a wave of locusts, almost brought about the economic collapse of the Kingdom of the Franks in Jerusalem.

ELTON[31] recently analyzed the cyclic fluctuations of voles. At least two types of cycles are evident in every country, for none of which there are clear causes. They certainly have nothing to do with the sunspot cycles. The most recent theories, both connecting the sensitive period of sexual maturation, refer if either to cyclic changes in ultraviolet radiation (SHELFORD[32]) or the recurrent fluctuations of the short lunar cycles (SIIVONEN[33]).

The briefest cycles are those of 3.6 year intervals. These are conspicuous in northern latitudes and as far south as Israel. In Southern Russia and in Israel cycles of ten-year intervals are much more accentuated than the shorter ones (Fig. 20). Our last peaks were in 1930, 1940 and 1950. In addition to these two types of cycles, every century witnesses two or three very large outbreaks and we have not yet ascertained whether these have a regular or irregular occurrence. The last flooding of the entire Levant with such a catastrophic vole outbreak was in 1930/31[34].

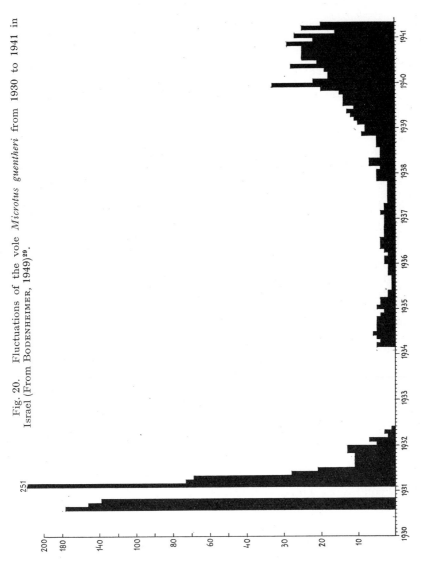

Fig. 20. Fluctuations of the vole *Microtus guentheri* from 1930 to 1941 in Israel (From BODENHEIMER, 1949)[29].

During our thirty years of continuous analysis of these vole outbreaks in Israel, the influence of climatic factor fluctuations was considered at the outset – temperature, humidity, quantity of food, density of population – with their effect upon the fecundity and mortality of the voles. It became obvious that voles avoid extreme injurious intensities of these factors with facility, by means of horizontal or vertical migrations (within the soil). HAMILTON[35] pointed out this same phenomenon in the U.S.A. which we observed in Israel. At the beginning of an outbreak, the fecundity of the female is almost doubled. Instead of 3 to 7 young per litter there are 6 to 14 in the pregnant female, and the percentage of pregnant females is much higher than usual. This would indicate the action of a gonadotropic substance stimulating the number of eggs produced and sexual activity or, at least, efficiency. Our earlier experiments with prolan, the gonadotropic hormone of the pituatary anterior lobe, administered subcutaneously or orally, had produced precisely that effect – almost doubled the fertility by higher litters and shorter intervals between subsequent pregnancies of the same female[36]. But what induced this sudden hormonal activity, accompanied by higher fertility? We have already stated that no single climatic factor can be connected with this rise of reproductive physiology, which also exceeds by far the normal seasonal peak of gonadotropic seasonal activity. At the time, FRIEDMAN discovered a substance with gonadotropic effect which appeared at intervals for a short season simultaneously in a number of agricultural crops[37]. This substance caused production within the female rabbit (which ordinarily enters the oestrous cycle only after copulatory stimulation) even without the presence of males and without the copulatory act. BRADBURY[38] revealed, however, that this substance has nothing in common with the gonadotropic substance of the pituitary but is some neurotoxic substance stimulating the pituitary similarly. This led to a theory ascribing the vole outbreaks to feeding on plants producing periodically a substance with a gonadotropic effect. This represents, of course, no final solution – it merely refers cyclic vole outbreaks to cyclic changes in plant physiology concerning the nature of which we are still wholly ignorant.

The statistical analysis of European *Microtus agrestis* populations by LESLIE & RANSON[39] opens the way to comparison with that of our Levant vole. The European vole has a life-expectation of 208 days, with a maximum empirical longevity of 693 days under laboratory conditions. The average female has 5 or 6 pregnancies with 3.5 to 4 young per litter. The most fertile age of the mother is from 80 to 200 days. The average physiological fertility of the female is thus about 20 young.

Under these conditions of fertility and mortality, the population may increase tenfold within six months. This is in the laboratory, under optimal conditions. In nature, however, there is no reproduction in winter, thus leaving a maximum period of 182 days for reproduction during the year. The mortality of a stable vole population is about 50% during the non-reproductive season. Thus the fluctuations of the onset of spring – and accordingly of those the climate between March and June – are of very great importance in nature. A late spring induces not only the retardation of reproduc-

## Table XVI.

Fertility of *Microtus guentheri* on irrigated and non-irrigated soils in Emek Yezreel 1947–51

| Month | On irrigated soil | | | On non-irrigated soil | | |
|---|---|---|---|---|---|---|
| | Females trapped per month | | Embryos per pregnant female | Females trapped per month | | Embryos per pregnant female |
| | non-pregnant | pregnant | | non-pregnant | pregnant | |
| 1947 | | | | | | |
| XI | 12 | 4 | 4.5 | 9 | 2 | 5.5 |
| XII | 7 | 7 | 7.2 | 5 | 10 | 5.2 |
| 1948 | | | | | | |
| I | 8 | 11 | 8.5 | 25 | 5 | 4.6 |
| II | 3 | 9 | 10.0 | 13 | 1 | 8.1 |
| III | – | – | – | 8 | 21 | 9.2 |
| IV | – | – | – | 18 | 21 | 9.2 |
| V | – | – | – | 19 | 13 | 8.8 |
| VI | 14 | 2 | 5.0 | 6 | – | – |
| VII | – | – | – | 15 | – | – |
| VIII | – | – | – | 4 | – | – |
| IX | – | – | – | 2 | – | – |
| 1949 | | | | | | |
| III | 8 | 19 | 11.1 | – | – | – |
| IV | 2 | 4 | 11.0 | 5 | – | – |
| V | 1 | 2 | 10.0 | 16 | – | – |
| VII | 4 | – | – | 13 | – | – |
| VIII | 10 | 2 | 5.5 | 11 | – | – |
| IX | 7 | 12 | 6.1 | 9 | – | – |
| X | 2 | 7 | 6.7 | 10 | – | – |
| XI | 9 | 6 | 6.3 | 10 | – | – |
| XII | 9 | 2 | 9.5 | 8 | 2 | 7.0 |
| 1950 | | | | | | |
| I | 10 | 5 | 9.5 | 4 | 1 | – |
| II | 3 | 8 | 10.5 | 2 | 3 | 9.0 |
| III | 7 | 1 | 11.0 | 4 | 6 | 10.0 |
| IV | 3 | 8 | 12.0 | 8 | – | – |
| V | 13 | 6 | 6.7 | 1 | – | – |
| VI | 10 | – | – | – | – | – |
| VII | 10 | – | – | 14 | – | – |
| VIII | 10 | – | – | 4 | – | – |
| IX | 9 | – | – | 5 | – | – |
| X | – | 4 | 7.0 | 5 | – | – |
| XI | 1 | 9 | 7.5 | 1 | 3 | 4.7 |
| XII | 3 | 12 | 10.0 | 3 | – | – |
| 1951 | | | | | | |
| I | 3 | 9 | 7.4 | 1 | – | – |
| II | 2 | 12 | 10.1 | – | – | – |
| III | 1 | 12 | 11.6 | – | – | – |
| IV | 5 | 7 | 7.0 | – | – | – |
| V | 13 | – | – | – | – | – |

tion but also increases the winter mortality of the voles. A very favourable early onset of spring, on the other hand, may lead eventually to an outbreak. The European vole constantly exists on the verge of extermination.

The corresponding data for the Levant vole are a life expectation of 718 days, the maximum longevity observed being 1900 days, under laboratory conditions. The average female has 13 pregnancies with 5 or 6 young per litter, an average total of 72 young per female.

Both species show, in a stable population, an increase and mortality of 12.4 ‰ per day. The Levant vole too has in nature a prolonged period of non-reproduction, but this occurs during the dry summer months. These data coincide with the type of population increase and of outbreaks all over the area, meaning that the motor for increase seems to be the initial rise of fertility.

For a very long time we remained in total ignorance of the factors causing the rapid breakdown of the outbreaks. We sought first the factors of natural decimation. As no climatic factors are directly responsible for the cessation of vole outbreaks, we looked for destructive factors commonly considered important – enemies and disease. Our investigations revealed – in agreement with those of ERRINGTON, ELTON and others on predation, that this is a minor factor, reducing the interest but not the capital of a population. It is easy to demonstrate that its influence is much lower at the peak of an outbreak than at normal population levels.

The long and patient research of ELTON, others and ourselves have definitely indicated that local epidemics of mice-typhoid, *Toxoplasma* and others may occur, but not one case is known where an entire area has been cleared of voles by any disease. Epidemics are certainly not the usual mechanism to end vole outbreaks.

On the other hand, if fertility is reduced to half the norm, there is inevitably a rapid and enormous reduction, just as the doubling of fertility brings about quick population increase at the outset of each outbreak.

The two years – 1948 to 1950 – served to give us the mechanism of the reduction of an outbreak. These were years of drought. Reproduction was very low to nil for seven to nine months, even in irrigated areas. Voles became almost extinct over all areas during these two successive periods of drought, except for the natural micro-areas of favourable habitat, such as swamp margins, railway and road margins, irrigation channels, etc. (Table XVI). It thus became clear, to our great surprise, that fertility is also the key to the decline of outbreaks[40].

We should here add that our animal has an ecological life-expectation of 70 days in nature (in contrast to the protected laboratory populations), of which about 5% only survive in nature half a year, even less a full year (Table XVII). Under these conditions the number of survivors must be reduced rapidly and catastrophically after a period of sterility during 7 to 9 months in summer. If such periods occur in two successive years we should not be astonished if the dense outbreak population is replaced by a vole-desert over wide areas.

**Table XVII.**
Age Distribution of Voles Devoured by *Tyto alba*

| Weight of Lower jaw | 80 | 120 | 160 | 200 | 240 | 280 | 320 | mg |
|---|---|---|---|---|---|---|---|---|
| corresponding age | 25 | 35 | 50 | 70 | 100 | 150 | 180 | days |
| per cent of voles | 2 | 9 | 23 | 31 | 18 | 11 | 5 | % |

This section has been called "Vole populations in Israel" since parallel research on voles in England[41] and Germany[42] led to quite different conclusions. Other species, principally *M. agrestis*, are involved in these countries and, as the author has had no opportunity to study these on the spot, he cannot offer a definite view on results from Northern Europe. The latter also describe a clearly outbreak-crash type of population movement. The assumption there is that of a purely biological cycle. During the outbreak a profound physiological disturbance of the metabolism and behaviour of these highly asocial voles is supposed to develop, speedily leading to partial heavy sterility not only of the outbreak generation but also of its offspring. This reduced fertility to sterility of the parent and first filial generations are indicated by missing or almost absent age-classes in the structure of the vole population, apparently connected with overdeveloped adrenal cortex and spleen.

The deep physiological disturbance is manifest also in reduced weight for every age-class and in heavy intraspecific strife[41]. Concerning the latter symptom, we have just published a prolonged series of *M. guentheri* breedings in large cages, indicating clearly that no prolonged stress occurs due to this factor, while brief periods of internecine activity take place at long intervals. This, however, is under the unnatural condition of large cages – we have no indication that similar periods of amok occur in nature, where any serious struggle can almost always be avoided by abandoning the area[43].

We fail to understand a mechanism which would be one of reproductive disturbance, followed some generations later by relaxation, leading rapidly to a fresh outbreak. From all the data hitherto published it seems clear that the dynamics of vole populations in the north differ considerably from those of the Levant vole. Yet even at this stage we may be permitted the following comments:

(1) The entire compensation mechanism of the individual or the population has not been considered. Detailed studies in this

direction will no doubt alter the presentation of the case considerably.

(2) We cannot conceive that profound physiological disturbances and their manifestations in reproduction should be transmitted to a second or third generation living in an environment where intraspecific strife is practically absent in a sparse and scattered population. It should be noted that the medical theory of stress effect on human beings varies greatly from that put forward here for voles. In man it is a transient "stress" of short duration which produces the shock. Corresponding studies on the effect of long-lasting "stress" are still under observation in man, but this definitely has very different symptoms from those of short stress.

While this book was going to press a clear description of the European *Microtus agrestis* cycles was published by F. FRANK[43a]. We wish to give its essence here.

Great importance is given to the high reproduction potential, expressed by the high weight of the litters, the suckling maturity and the long season of reproduction. This is essentially, with minor variations, true for all voles. We do not believe, however, in view of our experience, that the reproductive maturity of the still suckling young is a general phenomenon.

Equally important is a high carrying capacity of the environment, which includes the quantity of food, cover, sunlight (or daylight) good overwintering places, favourable groundwater conditions. In all areas of outbreaks we find – despite different geological, etc. situations – the same type of monotonous, open, cultivated steppes. FRANK stresses that true cycles appear only when the carrying capacity is high. In Central Europe the cycles are released by cultivation. Control by ecological change of this type of habitat is recommended.

A third series of factors is called by FRANK "condensation potential". The degree of population increase depends upon certain mechanisms of behaviour which enable many cyclic species to co-exist at an uncommonly high population density. Among the many points which are enumerated under the condensation potential we may mention the contracting territories when the population grows abundant; an increased sociability in winter, which serves as protection against the loss of body heat; the "common nest" of the "great family" at high population contraction; the mechanism of higher male mortality; etc.

"Because this rapid population increase cannot be regulated by normal mortality and dispersal, more efficient regulatory mechanisms are called into play. When the supportable density of population is approached, restriction of reproduction and accelerated individual emigration take place, but these are not enough to keep the population within the limits set by the carrying capacity of the environment. When supportable density is exceeded, crash, caused by shock disease, occurs in the following winter. Psychological stresses (such as crowding and competition) and physical stresses (such as food shortage) produce a "readiness" for crash, but the real trigger is largely the additional meteorological stress of winter. Three years are ordinarily required to reach this point, hence an autonomous and strict 3-year periodicity exists. The seldom-occurring deviations ("shifting of phase") are caused by uncommon meteorological conditions. Unusually severe winter synchronize the periodicity of isolated populations over large districts.

"On the whole, cycles take place where the high biotic potential of the species is fully realizable in the optimal biotopes of plague or other cyclic districts. It seems remarkable that this disproportion is not balanced by selection. Cycles have undoubtedly gone on from time immemorial, and the quick succession of generations of voles should have favoured such an adaption in a relatively short time[43b]".

It is well that we have now an organized theory upon which to work. The depressing influence on the following generations is reduced to the after-effects of the difficulties in bringing up the young of the first filial generation by the weakened parents. This is great progress, as is the entire mechanism evolved by FRANK, which comes much nearer to what a final analysis of the European vole cycles might accept.

## 6. The Muskrat populations of Iowa

This section could not be concluded without mention of the muskrat populations of Iowa. In this case we have had no prolonged personal acquaintance with the animal and its population problems, yet the lengthy and mature investigations of P. A. ERRINGTON on the muskrat in Iowa (U.S.A.) are of such far-reaching and fundamental importance that at least an epitome, quoting ERRINGTON's own words principally, must be given here. They concern two principles in the main:
   (1) The oft-misunderstood and exaggerated role of predation in nature; and
   (2) The complicated mechanisms of compensation, interplay, interruptions, deflections and successions that characterize free-living populations[44].

Neither of these principles is novel, but their interpretation, based upon ERRINGTON's vast experience and their exposition, is a definite contribution to the history of ecology.

PAUL ERRINGTON is today the outstanding authority on all problems connected with predation, especially in mammals. His main conclusions were contained in an early paper on one of the most efficient predators, the Great Horned Owl *(Bubo virginianus)* in North-Central United States[45]. Therein he expressed the doubt that this owl and associated predators exerted a dominant influence upon their prey populations. Even great concentrations of predators seldom utilized more than a small proportion of the rabbits and mice which form their staple food and are conveniently available. As a rule, they turn to other mammals or birds when the latter temporarily become more available than the prey staples. Overpopulation of habitat by non-staple prey species has been accompanied by some of the most pronounced rises in representation of these types in the diets of such flesh-eaters as the Horned Owl, crises precipitated by weather, destruction of environment, human activities, etc. were often reflected, as well, by the response of predators to increased vulnerability of given prey animals. This general conclusion – that predation has no decisive influence on the populations of vertebrates – is repeated by ERRINGTON in a comprehensive review[46].

It was already evident, in this study on the Horned Owl, that the predator-prey relationships were flexible and that predators did not always respond to apparent change in the availability of prey.

"Nevertheless, unrelieved basic insecurities of prey populations were often attended by response of some predators sufficient to compensate, at least in part, for lack of utilization by others; and compensatory tendencies in loss rates of prey animals were noted under a wide range of conditions having little evident connection with kinds and densities of predators."[47]

ERRINGTONS's mature analysis of Mink *(Mustela vison)* predation upon Muskrats *(Ondatra zibethicus)* in Iowa is one of the classics of recent research. In his summary we read[48]:

"It was found that strangeness of environment, intra-specific intolerance, and drought were especially important in predisposing muskrats to predation, as by minks; and that kinds and numbers of wild predators doing the preying, with a few apparent exceptions, had little bearing upon the net mortality suffered by the muskrats. For reasons of intercompensatory (automatically adjusting) trends existing in reproductive as well as loss rates of the muskrats, severity of predation may leave highly misleading impressions as to population effect. Even when locally nearly annihilative, predation rarely showed evidence of functioning as a true population depressant, insofar as it usually only took the place of some other mortality factor and as, in its absence, some other factor, particularly intraspecific strife, tended in its turn to become sufficiently operative to compensate for decreased predator pressure."

This becomes clearer when studying the seasonal aspects of mink predation upon muskrats in the area under ERRINGTON's survey:

Remains of mature or maturing muskrats may often, if not usually, be conspicuous about marshy mink retreats in spring (April/May) during the spring dispersal from wintering to breeding quarters. But the evidence is that the victims are chiefly surplus, unmated, strife-battered, wandering males – a part of the population that is typically lost in many localities, irrespective of whether through predatory enemies, intraspecific strife or other factors.

Adults of the breeding populations established in family territories suffered slight loss from natural enemies unless placed under overwhelming disadvantage by drought emergencies or the like; even on dry marshes hunted over by minks and foxes, some old muskrats were able to live on and on during the warmer months.

The young, immature muskrats suffered most from depradations of minks when drought conditions were acute and when overpopulation prevailed in their habitat. Only predation of muskrat litters close to mink dens, concerning only a small part of the area, did not show intercompensation. The heaviest losses of young accompanied rather than governed the directions taken by population curves. If populations of young were basically insecure, their losses tended to occur through several agencies, single or combined. When mink predation was severe, other losses had a tendency to diminish in proportion; lessening of mortality through predation was largely counterbalanced by increased killing of young by older muskrats or by acceleration of miscellaneous losses. Apart from human, climatic and general environmental factors relating to recovery of muskrats during the breeding season, the species' own psychology rather than its theoretical reproductive potential, or the impacts of its enemies, seemed to be the primary regulator of local rates of increase. Adults under crowding not only turned deadly teeth upon annoying young but were also known to stop breeding ahead of schedule; conversely, one of the main antecedents of breeding prolonged late in the season was poor success in the rearing of earlier litters.

The interval between the end of the breeding season in late July and August and the hard frosts of late September and October was one of minimal intraspecific friction in muskrat habitats. Mink predation was light at this season except that upon drought exposed muskrats and upon those which for any reason had fallen into hazardous wanderings.

Vulnerability to minks of the drought-exposed and the transient muskrats was pronounced during cold weather but such animals were from one cause or another so inevitably doomed – actually dying more from intraspecific attack, hunger, and cold than from predation – that the exact medium of their elimination in the end made scant difference, biological or otherwise. Seldom could it be judged that absence of minks or of other predators would have meant greater winter survival of patently handicapped individual muskrats. There was, nevertheless, conspicuous mink pressure upon some populations that were apparently wintering with fair security except from this cause, and, to a large extent, such predation would seem to be noncompensatory or of intercompensatory action delayed until the spring dispersal, perhaps linked with increasing irritability and unrest.

"On the whole, any factor, whether of "natural" origin or associated with man, that promoted instability in muskrat populations heightened vulnerability to predators, and the resulting losses were likely to show a high degree of intercompensation ... Mink predation, in contrast with drought, does not have the attributes of a highly selective agency, despite its representing one of the most severe pressures exerted by a vertebrate enemy upon a prolific rodent."

The applications to fur management in that area (ERRINGTON[49] p. 802) are of interest. The data suggest that little increase in revenue from muskrats would be gained by deliberate repression of minks in this area, where the muskrat pelts are taken in autumn and winter.

Where muskrat trapping is done in early spring, it is probable that the reduction of minks by mid-winter would serve more or less to bolster top-heavy muskrat populations long enough to permit utilization by man. In either event, considering the characteristics of the various types, mink predation occurs and may be severe; the management and harvest of both minks and muskrats on a sustained yield basis should be preferable in fur economics to intentional sacrifice of one resource for the sake of limited or dubious benefit of the other.

Most important are ERRINGTON's conclusions[49] (p. 299):

"I question that the biotic potential of my Iowa muskrats could be expressed numerically with even a dependable prospect of 50 per cent accuracy. Not only may the average sizes of their litters vary with "cyclic" phases and intraspecific density, etc., but the numbers of litters born per adult female in a breeding season also depend upon degrees of crowding and the properties of the habitat. A population of breeding females may conceive or give birth to twice as many young as many other populations of like densities and environmental status if the latter population merely happened to have been more fortunate in rearing its early-born litters. One may be reminded of the prompt renesting of many species of birds following destruction or abandonment of early clutches of eggs.

"In another part of a balance sheet equation, we have losses from a variety of factors. Some are density-dependent and some are not; some are intercompensatory and some are not; but a substantial proportion in the muskrat and all other vertebrates with which I have worked are both density-dependent and intercompensatory. A usually fallacious procedure is to assign values (arbitrarily or on the basis of research findings) to the different agencies of loss, to add up the values, and to consider the total arrived at as a figure reflecting the workings of Nature. Through this kind of calculating, bad distortions may be dignified by a semblance of scientific methodology that might deceive anyone not familiar with its analytical pitfalls. Unless they consider the automatic shifts taking place within life equations, well may anyone wonder that many thriving species can exist at all in the face of the enemies they have or the losses they suffer!

"First, we must not forget that extraordinary mortality may offset itself by stimulating reproduction, more young being produced as a result of more dying. Likewise, that the total amount of loss suffered by a given population in a given time and place depends in part upon whether or not more individuals have been produced than the available habitat can securely accommodate.

"Excess populations for an area may be numerically large or numerically small; they may comprise a large or a small proportion of the total number of individuals present. Sometimes, there may be no real excess population for an area, even for an abundant species. In marginal habitats, there may be no real security for a particular species, even when it may be barely represented. The "normal" range of variations in this respect is such as to warn us against improper assumptions of concreteness. A given muskrat population may appear to be remarkably secure at an average density of 15 per acre; another population may attain that level but be so top-heavy that three-quarters are candidates for elimination in one way or another; another may not be able to maintain itself at 15 per sq. mile, though temporary densities greatly exceeding this figure may occur.

"The case histories of muskrat populations bespeak much regulation by formula, not just a formula in which values for the terms conveniently "stay put" so much as one operating along a sliding scale as populations rise or decline within the framework of toleration limits and environmental requirements of the species, "cyclic" phases, etc. Except in the sense that everything living must ultimately die, life equations of the muskrats tend to be replete with flexibilities.

"A value assigned to a mortality factor that merely substitutes for another should not be considered the equal in population significance with one that genuinely brings about a net lowering of the population. What difference does it make biologically if a mink or a fox or another muskrat killed a young muskrat as long as death of that individual essentially made it possible for another young muskrat to continue living? The population impacts of a dominating emergency or epizootic are in an entirely different category. It is in mishandling of losses designated for specific agencies that one is apt to go farthest astray in balance sheet computations.

"We are still in explorative stages so far as concern natural checks and balances in animal populations, and the muskrat data give us scarcely more than indications as to what may be fundamental mechanisms. We can see that widely-held concepts as to limiting factors need a good deal of revision. The modern biometrician must be careful not to confuse with limiting factors those population by-products that occur incidental too, but with slight if any influence upon the population status of a species. He must be careful not to confuse the fact of losses with the effect of losses."

The study of a lethal and highly infectious haemorrhagic disease of uncertain causation[49] in muskrats yielded very striking results in the last decade. 13,000 mink scats were recorded during this decade from Central Iowa, 2415 of which contained remains of muskrats. 1344 of these pointed to scavenging of minks upon muskrats dead or dying from this haemorrhagic disease. Disease as a dominant factor, together with drought or freeze-out, accounted for 343 more muskrat-containing scats; 115 others had suffered from other diseases. All together, 1600 to 1700 of the muskrat-containing scats, or 65 to 70%, were due to the haemorrhagic epidemic and to other diseases. Somewhat less than 100 muskrats were killed in consequence of "intraspecific strife, old age, freezing, drowning, shooting and trapping, dog and fox predation, highway traffic" and rarer mishaps.

The remaining 674 scats (28%) are divided by ERRINGTON into seven categories of mink predation: 46 resulted from predation upon drought-exposed muskrats, 78 from predation at times of freeze-out (of healthy individuals), 43 individuals were linked with storm evictions and floods; general environmental deterioration (e.g. after killing of better food plants by high water) accounted for 150 scats. The two last categories concerned largely very young muskrats "which are consistently among the best candidates for elimination by predators, when something goes wrong at a marsh."

22 scats were killed with the autumn migration and 62 with the spring migration. "Most of the predation upon autumn wanderers was centred upon a wretched, strife-torn overflow from deteriorated habitats." Increasing restlessness and friction in crowded habitats were responsible for 273 scats, largely concerning individuals with defects, less vitality or sexually active individuals, mainly males. The remaining 40 scats might be due to chance encounters of minks with muskrats, especially with young individuals.

Only 28% of the 2415 muskrat-containing scats were due to predation, and this under a great variety of conditions. They concerned only a very few individuals old enough to take care of themselves and in full health and vigour.

With regard to selection pressure of predation, ERRINGTON is of the opinion that when both species met first, it was probably great. At present it is negligible.

"There are, in the literature, some instructive case histories of native faunas finding themselves confronted by predators with which they have had little or no prior racial experience. Where the new predator has been formidable enough or the prey vulnerable enough, spectacular, or even annihilative, losses from predation have resulted. One case that I often think of relates to Jamaica, where[50] the black-bellied tree duck *(Dendrocygna autumnalis)* nested on the ground and in open fields before the introduction of the mongoose *(Herpestes griseus)*. The mongoose almost exterminated the tree duck in a few years, but a remnant survived because of safer nesting habits and then began a gradual increase in numbers. This is the kind of adjustment that I suspect has taken place in the early years of many predator-prey relationships that now have the aspects of permanent stability. My view is that this is about the way that evolution may leave some of its biggest imprints on the muskrats as a prey species, great tests that a prey species must pass at rare intervals, followed by generation after generation of the prey fully endowed with the hereditary equipment to enable the most commonplace of individuals to do enough of the right thing when needing to."

Of the many more ecological field studies of vertebrates we would like to recommend a few very heartily as introduction into animal ecology and its problems, when it is disrobed of all tables, ciphers and formulas. There is the delightful description of the mackerel's varying history by RACHEL CARSON[51], a real revelation. For bird lovers the selection is bigger. We recommend beginning with books on the Bobwhite Quail[52], the Robin[53], or the Fulmar[54]. We mention these in order to indicate that any individual life-history of any animal species is the very best introduction to ecology available, and such literature should be warmly recommended to every beginner or lover of animal ecology.

# IV
# BIOLOGICAL EQUILIBRIUM IN NATURE AND BIOLOGICAL CONTROL

## 1. The Case for Biological Equilibrium

The role played by biotic factors is one of the great problems of ecology. Although food, population density, diseases, and the reaction basis of the organism also belong to the biotic factors, we shall here discuss principally the influence of enemies upon their host or prey, which – just as temperature amongst the abiotic factors – has best been studied.

In the first period of our knowledge enemies were regarded as the primary factor regulating animal populations. This opinion found its classical expression with regard to insects in the report of HOWARD & FISKE[1] on the importation of the parasites of the Gipsy-Moth and the Brown-Tail Moth into the United States:

"To put it dogmatically, each species of insect in a country where the conditions are settled is subjected to a certain fixed average percentage of parasitism which in the vast majority of instances and in connection with numerous other controlling agencies, results in the maintenance of a perfect balance. The insect neither increases to such abundance as to be affected by disease or checked from further multiplication through lack of food, nor does it become extinct, but throughout maintains a degree of abundance in relation to other species existing in the same vicinity, which, when averaged for a long series of years, is constant.

"In order that this balance may exist it is necessary that among the factors which work together in restricting the multiplication of the species there shall be at least one, if not more, which is what is here termed facultative, and which, by exerting a restraining influence which is relatively more effective when other conditions favour undue increase, serves to prevent it.

"A very large proportion of the controlling agencies, such as the destruction . . . by (extreme) climatic conditions, is to be classed as catastrophic, since they are wholly independent in their activities upon whether the insect which incidentally suffers is rare or abundant. The average percentage of destruction remains the same, no matter how abundant or how near to extinction the insect may have become.

"Destruction through . . . birds and other predators, works in a radically different manner. These predators are not directly affected by the abundance or scarcity of any single item in their varied menu. Like all other creatures they are forced to maintain relatively constant abundance among the other forms of animal and plant life, and since their abundance from year to year is not influenced by the abundance or scarcity of any particular ... prey they cannot be ranked as elements in the facultative control of such species. On the contrary, they average to destroy a certain gross number of individuals each year, and since this destruction is either constant, or, if variable, is not correlated in its variations to the fluctuations in abundance of the insect preyed upon, it would most probably represent a heavier percentage when that insect was scarce than when it is common ... A natural balance can only be maintained through the operation of facultative agencies which effect the destruction of a greater proportionate number of individuals as the insect in question increases in abundance. Of these facultative agencies parasitism appears to be the most subtle in its action. Disease ... or insufficient food supply does not as a rule become effective until the insect has increased to far beyond its average abundance."

This is the classical plaidoyer for biological equilibrium.

## 2. The Case for Climatic Control

The concepts just described dominated ecological thought during many decades until a reaction set in, stressing the fact that abiotic factors, principally climatic, control the numbers of individuals. Long series of observations, mainly on insects, showed an unexpected degree of parallelism between fluctuations of certain weather facts as temperature, precipitation, humidity, etc., and those of many animal species. Similar fluctuations are observed in such insects and other animals among which enemies and diseases certainly do not play an important role at any period of the gradation.* Moreover, it has been demonstrated by introducing physiological laboratory methods of research that climatic factors, especially temperature and humidity, exert a dominating influence on the vitality of poikilothermous animals at least, and that such processes as fertility, egg-production, longevity, etc. are influenced even by relatively small deviations from the optimal combinations of these factors, the vital optimum. These weather influences are of dominant importance during certain, generally short, periods of the life-cycle, which are the most sensitive ones[2].

Statistical analysis of gradations showed that, as a rule, the share of destruction by weather factors is much larger than that of all other components, and it was therefore concluded that climate is the main controlling factor of numeric abundance.

---

* In studying animal populations we define a gradation as the interval between one lowest point of the population ebb and the following low, thus including one full wave of the fluctuation.

These conclusions have been well presented by Uvarov[3]:

"The theory that all living organisms are in a stable equilibrium so far as their relative numbers are concerned is widely recognized to-day. It is however in direct contradiction to the facts. While it is true that an increase in numbers of a species is usually only temporary, and that a decrease will, sooner or later, follow, there are no proofs that the fluctuations in the two directions are of an approximately equal magnitude, as in the case of a pendulum. Nor is the so-called normal number a constant . . .

"The theory of stable equilibrium is based on the assumption that the numbers of an organism depend mainly on the numbers of their enemies and on the quantity of food, i.e. on factors which in their turn are dependent on other organisms. No one will deny the controlling value of these factors, but the evidence collected should go far towards proving that the key to the problem of balance in nature is to be looked for in the influence of climatic factors on living organisms. These factors cause a regular elimination of an enormous percentage of individuals under so-called normal conditions, which in fact are such that insects survive them, not because they are perfectly adapted to them, but only owing to their often fantastically high reproductive abilities. Any temporary deviations in the climatic factors, however slight they may be, affect the percentage of survival, either directly, or indirectly (through natural enemies and food-plants), and thus influence abundance. Further, the ecoclimatic succession and the geological succession of climates favour the survival and the relative abundance of some species, and condemn others to a complete extermination. Of course, the food, the natural enemies, the competition between individuals and between species, all play their part in these evolutionary processes, but none of these factors are independent and primary in character, since they themselves are deeply affected by climate. It is possible to imagine an insect with no natural enemies and without any need to compete for food, shelter, etc. and there is no doubt that this, if not very commonly, does actually happen in nature, but an insect living under natural conditions and yet free from climatic influences is an absurdity . . . To study the dynamic balance of a species without taking into consideration its climatic environment, is an utterly unscientific procedure."

This is the case in nuce for the climatic school.

## 3. Mathematical and Experimental Approach

The problem of intra- as well as interspecific regulation has been approached within the past decades from the experimental as well as from the theoretical mathematical point of view.

The problem of intraspecific regulation has been discussed amply in a previous chapter[4]. The logistic growth of the population is the result of given conditions. Within a limited space a population is being more and more forced from an initial exponential growth into a logistic one, by population pressure, i.e. by a factor determined mainly by the reaction basis of the organism itself.

The problem of interspecific regulation has been attacked by theoretical analysis by VOLTERRA[5] and LOTKA[6] and experimentally by GAUSE[7].

Mathematical analysis must necessarily begin with a simplified hypothesis in order to obtain general conclusions. VOLTERRA based his analysis on the assumption that only the voracity of the predator with regard to a given prey and the reproductive capacity of those species living together were the effective factors, all other conditions

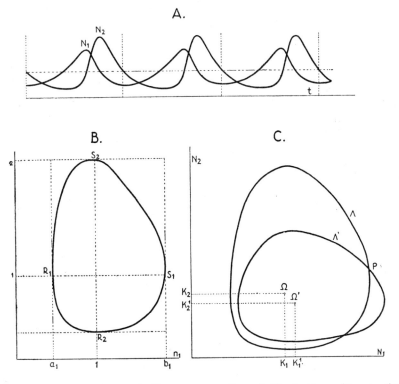

Fig. 21. Illustrations to the VOLTERRA equations: A. Fluctuations of two species one of which devours the other. B. Cycle of numeric variation in two species, one of which devours the other; each point of the cycle indicates the numbers of the predator and the prey living at a given moment. C. Changes in the diagram B when both species are uniformly and proportionally destroyed (3rd law of VOLTERRA).

being constant. On this basis he arrived at the following fundamental laws governing the fluctuations of two species living together:

(I) "Law of the periodic cycle. The fluctuations of the two species are periodic and the period depends only upon 1, 2, and C (namely upon the coefficients of increase and decrease and the initial conditions).

(II) "Law of the conservation of the averages. The averages of the numbers of individuals of the two species are constant whatever may be the initial values of the numbers of individuals of the two species just so long as the coefficients of increase and decrease of the two species and those of protection and of offence (1, 2, 1, 2) remain constant.

(III) "Law of the disturbance of the averages. If an attempt is made to destroy the individuals of the two species uniformly and in proportion to their number, the average of the number of individuals of the species that is eaten increases, and that of the individuals of the species feeding upon the other diminishes. Increase of the protection of the species fed upon increases, however, both the averages[5]".

Other types of "associations" as well as many special cases have been discussed. However, of fundamental importance is the fact that VOLTERRA and his followers obtained fluctuations, based on the above-mentioned assumptions, which are strictly in accordance with the theory of HOWARD & FISKE (Fig. 21), but only concerning host-parasite fluctuations, not concerning the general interspecies relations of an area.

Experiments on such and similar interspecific influence of population growth have been made and mathematically analysed by GAUSE. Some extremely illuminating cases may be summarized:

(1) Two species competing in identical niches. Two ciliates may grow under identical conditions within a test-tube. One species, *Glaucoma scintillans*, fits better into the environment, and in concordance with VOLTERRA's calculations the second species, *Paramaecium caudatum*, is expelled finally from this environment, if both grow in a mixed culture.

(2) Two species of different niches, but in competition for the same food, *Paramaecium caudatum* and *P. bursaria*, feeding on *Saccharomyces exiguus* and *Bacillus pyocyaneus*. Each species having the advantage of its proper niche, both species may coexist for an infinite period, which result is again in agreement with biomathematics.

(3) Two species, one of which feeds on the second. Fig. 22 shows the consequence of the introduction of the predatory *Didinium nasutum* within a population of its prey *Paramaecium caudatum*. The possibilities are more complex than in the earlier-mentioned cases. If the test-tube is without oat sediment, the predator destroys

its host in this extremely simple and limited environment fully, and has to die from starvation in consequence (Fig. 22, A, a). But when this microhabitat contains a refuge for the prey, e.g. in form of an oat sediment, and if *Paramaecium* and *Didinium* are simultaneously introduced into the microhabitat, the number of predators increases somewhat and they devour a certain number of *Paramaecia*, but a considerable amount of the prey is in the refuge and the predators cannot attain it. Finally, the predators die out entirely owing to lack of food, and then an intense multiplication of the *Paramaecia* (no cysts of *Didinium* being observed) begins in the microcosm (Fig. 22, A, b). An experiment under more complex conditions and more in conformity with natural conditions, showed the effect of regular immigrations of a predator into an environment with refuge (Fig. 22 A, c). At the first immigration the test-tube contained a few *Paramaecia* only, the predator died by starvation and the prey began to increase considerably. The second immigration met with a higher population level of the prey. *Didinium* increased considerably and *Paramaecium* was greatly reduced. The third immigration occurred at the population peak of *Didinium*

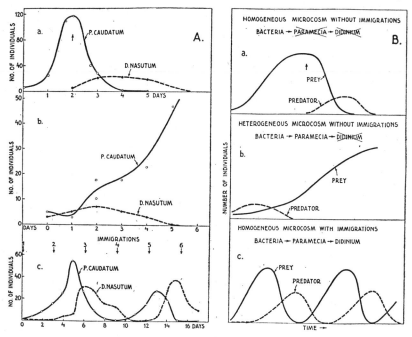

Fig. 22. Interaction of populations of two species, one of which *(Didinium)* preys on the other *(Paramaecium)*. A. Experimental results; B. Schematic curve. Explanations in text (From GAUSE).

and did not cause any essential changes. At the time of the fourth immigration the predator had already devoured all the prey, had become reduced in size and degenerated. The prey was now introduced into the test-tube, and a new cycle of growth of the prey population originated. Such periodic changes repeat themselves farther on. All these cases are bio-mathematical eventualities.

Considering the complexity of the phenomenon of animal fluctuations in nature it seems premature to apply the formulae of VOLTERRA to actual fluctuations observed. What basic mistakes may originate from such applications may be gathered from one very illuminating case. VOLTERRA's interest was aroused mainly by an interview with D'ANCONA, who observed that after interruption of the fisheries in the Adriatic during the first world war the percentage of selachians, i.e. of the main predators among the benthal fauna, rose greatly and declined parallel with the beginning of fishery after the war. This case induced VOLTERRA to attempt a mathematical analysis, and found its solution in the third law of fluctuation quoted above[8]. A complementary study made by the present author showed definitely that three such fluctuations of approximately equal size were observed between 1900 and 1930, and that the parallelism between interruption of fishery and increased percentage of selachians is doubtless only a chance one[9]. Fluctuations of hydrographical factors are probably responsible for them.

The objections to VOLTERRA's theory are many. V. A. BAILEY[10] proved that a steady state is impossible when the number of parasite species is lower than that of the host species. Even more unfounded is the theory that only an even number of interacting species can be in a steady state. This erroneous conclusion was made possible only by completely ignoring the changes in the age-structure of the population. He maintains that this error is by no means merely an accidental, aesthetical defect but rather a basic one of the theorem.

Another criticism of basic importance is by ANDREWARTHA & BIRCH[11] against the statement made by HUTCHINSON & DEEVEY – in agreement with VOLTERRA – that two species with the same niche requirements cannot form mixed steady-state populations in the same region, a statement often regarded as one of the chief foundations of modern ecology. They demonstrated that the formulae upon which these conclusions ar based are unsound.

A number of other experiments, apart from those of GAUSE, seem to confirm this conclusion. T. PARK[13a], for example, studied in a long-term project the interspecific competition of *Tribolium confusum* and *T. castaneum* under various constant conditions. While any of the species survives indefinitely under all experimental conditions when one species only is bred, the additional interspecific competition eliminates one of the two species always – the one which is less adapted to the particular situation – under certain conditions one, under others the one or the other of the two competitors surviving. *Callosobruchus chinensis* and *C. quadrimaculatus* in competition always ended with the elimination of the latter species[12].

All these experiments were carried out in a constant physical environment, and it is by no means certain that the same results would be obtained when the environment follows changes of nature. We may even quote a case[13], where the elimination did not take

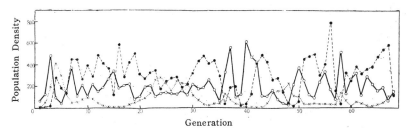

Fig. 23. Population fluctuation in the interacting populations of host and two parasites. Host: *Callosobruchus chinensis* (full line with hollow circles); parasite wasps: *Neocatolaccus mamezophagus* (dotted line with black dot) and *Heterospilus prosopidis* (dotted line with x). Over 70 generations all three species co-exist. (From UTIDA, S., Population Fluctuations in the System of Interaction between a Host and its Two Species of Parasite. Experimental Studies on Synparasitism. 3rd Report. Oyo – Kontyu. Vol. 11, no. 2 (1955) pp. 43—48).

place. UTIDA maintained a culture of *Callosobruchus chinensis* where two parasitic wasps of similar habits and niches, *Neocatalaccus mamezophagus* and *Heterospilus prosopidis* coexisted for three years in 70 generations. Fig. 23 demonstrates this fact as well as the alternating ups and downs of the wasp populations.

VOLTERRA's equations seemed to have solved the question of natural equilibrium, but nothing is less true than this conclusion. They have offered a valuable solution for the problem of why blind elimination by predators from a conglomerate of various populations produces longer and larger prey. This is the general fishery experience, e.g. that of the herring fisheries in the North Sea. Nearly all authors, including NICHOLSON, support the theorem that no eco-equal species can co-exist in the same niche for a prolonged time. Clear as the solution of this problem may seem, in thought, mathematical analysis, or even experiments under constant physical conditions, nature's problem cannot thus be solved. The sole solution is by prolonged observation under natural, fluctuating conditions. As there are no two exactly equiecous species, the answer will probably be that even very close species of similar ecological requirements will, under variable conditions, frequently have an alternating advantage over the competing species, and vice versa. Hence, both species may survive – or, if the advantage of one of the two species is really overwhelming, then one only of the two species may survive.

## 4. The Theorem of Nicholson & Bailey

Nicholson & Bailey[14] have certainly achieved the greatest progress in the mathematical presentation of the dynamics of animal populations. Their sound biological approach has a broad and reasonable foundation upon which their mathematical calculations are built, the attempt to find the agreeing principles of most bio-mathematical systems, the reiterated emphasis that many factors combined still retain their collective influences, the due consideration of the lag-effects caused by the development of animals just born as well as of the age-distribution of the population, and last but not least, the extended regular compensatory reactions in populations which enable them to survive and even to maintain surprising numbers in situations of discomfort and stress, etc.

Nicholson[15] has recently resumed the following fundamental biological assumptions which form the basis of his calculations, as follows:

"All biologists are aware that populations are influenced by many and varied factors, but an underlying orderliness is equally evident. The necessary first step in the investigation, therefore, was to classify the various types of factors according to their influence upon populations, and to define and study those pertinent facts about animals and their populations which are axiomatic, or so well established by generations of observers as to be unquestionably true. Of these the following established facts seemed to me to be most significant and to provide the basis for an understanding of population dynamics:

(1) "All animals have an innate ability to reproduce and to multiply under favourable conditions.
(2) "The favourability or otherwise of the environment for a given species determines whether its population is permitted to grow or is caused to decrease.
(3) "As every animal born must die, and can die only once, a population cannot persist for long periods in any given environment unless the birth-rate and death-rate are virtually equal (when each is averaged over a representative period); for if numbers of births and of deaths remained appreciably different the population would decrease or increase geometrically, so either directly falling to extinction, or causing its own extinction by overwhelming its environment, in a comparatively short time.
(4) "As populations grow, the constituent animals use up more and more of the limited available quantities of depletable requisites (such as food and favourable space), and increasing density often intensifies the action of inimical factors (for example, by increasing the densities of any natural enemies

dependent upon the animals concerned, or by increasing the concentration of harmfull metabolites).

(5) "Because animals produce such effects upon their environments, growing populations progressively reduce the favourability of some factors, whereas decreasing populations permit favourability to recover. Such compensatory reaction inevitably governs population densities at levels related to the properties of the animals and those of their environments.

(6) "Consequently, when operating in association with density governing factors, non-reactive factors (such as climate) may have a profound influence upon density, for many non-reactive factors influence either the properties of animals or environmental favourability.

(7) "Operating by themselves, however, non-reactive factors cannot determine population densities for, if sufficiently favourable, they permit indefinite multiplication or, if not, they cause populations to dwindle to extinction. On the other hand, they inevitably limit distribution to those areas within which they are favourable.

"These basic considerations are of general application, being quite independent of the nature and of the complexity or otherwise of the situations in which the animals live."

Most of these assumptions do not call for any discussion as they are accepted by all biologists. Paragraph 5 is usually correct and then of great importance. NICHOLSON[16] himself has performed a number of experiments which demonstrate this compensation principle very well.

In one series of breedings of the Sheep Blowfly *Lucilia cuprina* the population was limited by 50 g of larval food per day. Ground liver (except in I), water and sugar were supplied to the adults at all times in excess of their requirements.

Increasing destruction of emerging adults (f to h) led to the increased daily emergence of adults. This was caused by the reduced larval competition for food as a result of the fewer eggs laid (b) by the adults (g) reduced in number by the arbitrary destruction of a certain percentage of just-emerged flies, as the number of adults per unit of larval food increases with the decreasing larval density (h) (above a certain critical low level). Adult mortality of course also occurred naturally (j). The reduction of larval competition (1) increased the number of individuals (c) surviving the competitive stage, (2) it induced an increase in the size and vigour of the survivors, as expressed by the increase of the coefficient of fertility (m), the greater number of offspring (1) and the greater adult longevity (k).

"In the experiments compensation for imposed destruction was due to a combination of reduced larval competition, lessened direct

**Table XVII.**
Effects produced in *Lucilia cuprina* populations by various destructions and discomforts

### 1. Average of observed quantities

| Situation | % destruction of emerging adults (a) | Larvae hatched per day (b) | Adults emerged per day (c) | Adults destroyed per day (d) | Accessions of adults per day (e) | Natural deaths of adults per day (f) | Mean adult population (g) |
|---|---|---|---|---|---|---|---|
| E. No adults destroyed | 0% | 8794 | 157 | 0 | 157 | 157 | 1252 |
| F. 75% adults destroyed | 75% | 4121 | 192 | 144 | 48 | 48 | 464 |
| G. 95% adults destroyed | 95% | 1921 | 437 | 415 | 22 | 22 | 249 |
| H. 99% adults destroyed | 99% | 1330 | 909 | 900 | 9 | 9 | 128 |
| I. Bad oviposition site | 0% | 4261 | 234 | 0 | 234 | 235 | 1667 |
| J. 1 g. liver per day | 0% | 981 | 598 | 0 | 598 | 599 | 4042 |

### 2. Derived characteristics of the population

| Situation | % of larval competition (h) | total adult destruction (i) | mortality due to natural deaths of adults (j) | Mean adult life days (k) | Mean birth-rate per fly (l) | Mean coefficient of fertility (m) | Minimum coefficient of replacement (n) |
|---|---|---|---|---|---|---|---|
| E. | 98.2 | 0 | 1.8 | 7.9 | 7.0 | 55.8 | 1 |
| F. | 95.3 | 3.5 | 1.2 | 9.7 | 8.9 | 85.8 | 4 |
| G. | 77.3 | 21.6 | 1.1 | 11.4 | 7.7 | 88.0 | 20 |
| H. | 31.6 | 67.7 | 0.7 | 14.1 | 10.4 | 146.3 | 100 |
| I. | 94.5 | 0 | 5.5 | 7.1 | 2.6 | 18.2 | 1 |
| J. | 39.0 | 0 | 61.0 | 6.8 | 0.2 | 1.6 | 1 |

effects of adult crowding, and increased fertility, all these being reactive effects of the destruction."

All this is a good complement to the compensation effects of the muskrat populations discussed earlier.

Fig. 24 illustrates another experiment. The *Lucilia* population was limited by a daily supply of 0.5 g of ground liver for the adults at 25° C, the larval food being always in excess. The characteristic feature of the population was the maintenance of regular but violent oscillations in the number of adult flies. Significant egg-production only took place when the adult population was low, i.e. when the few surviving flies had a good chance to secure sufficient food to

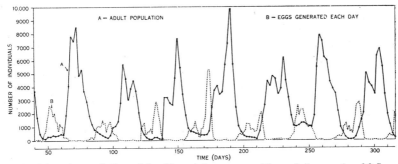

Fig. 24. A population of *Lucilia cuprina* governed by a daily supply of 0.5 g of ground liver for the adults, larval food being always abundant (From Nicholson, A. J., An Outline of the Dynamics of Animal Populations. Australian Journal of Zoology, Vol. 2, no. 1 (1955), pp. 9—65, fig. 3).

develop their eggs and lay them. The large fly populations thus always had to be reduced to small ones before the flies were able to obtain sufficient food for egg production. These deposed eggs gave rise to the new adult generation in about two weeks, and before the appearance of this new fly population the survivors of the earlier generation had dwindled away. The new cycle began with the sudden appearance of a large overcrowded fly population unable to obtain sufficient food for egg production.

"If increased acquisition of food were to cause fully mature adults to come into being immediately (instead of merely initiating the subsequent production of eggs and the still later development of adults) this prompt reaction would cause the system to be non-oscillatory. This is because reaction would cause the population first to approach, and then to maintain the equilibrium density of the species under the prevailing conditions, this being the density at which production of offspring precisely compensates for the loss of adults by death; for any departure from this level would immediately bring compensation reaction into play, and this would cease as soon as the equilibrium density was attained again. This, then, is the balancing mechanism which holds population density in general relation to the prevailing conditions, and the system of balance is often highly oscillatory, simply because animals commonly take a significant time to grow up, so causing a time lag between stimulus and reaction. During this lag-period the stimulus continues to generate more and more reaction, and this continues to come into operation for a similar lag-period after reaction has removed the stimulus."

NICHOLSON remarks with justice that such violent oscillations would usually suggest a periodic environmental factor in a natural population. In this stable experimental population, however, it was clearly based upon limitation of food supply in a peculiar stage. These oscillations are also more apparent than real: As we consider as population always all stages and ages of the population, the latter would actually approach a stable equilibrium. The restriction of a population census to the adult stage only is altogether unjustified. Hence NICHOLSON's remarks regarding stable population density are fully applicable to the present experiment.

While in this experiment the "average density" with 0.5 g of ground liver per day was 2520 flies, it was at 0.1 g of ground liver per day 527, i.e. almost exactly proportional to the limitation of the supply of the governing restriction of the adult food.

"Governing reaction does not merely operate to oppose any departures of a population from its equilibrium density, but also enables populations to adjust themselves to withstand very great environmental stresses (particularly when their inherent reproductive capacity is high) and to maintain themselves in a state of balance under widely different environmental conditions. Moreover, the reduction in density which adverse factors produce as a primary effect is always opposed by compensatory reaction, being lessened, or even converted into an increase in density, when the population adjusts itself to the continued operation of the adverse factor."

We have just given the basic biological assumptions of the mathematical theorem of NICHOLSON & BAILEY. A system of self-regulating populations is built up on these bases similar to that of Fig. 21B. If we conform to the principles of this book to avoid all mathematical formulae – as everything expressed in the precise language of mathematics can also be expressed in the usual language – we cannot here follow the trend of the mathematical equations of this system which appears the most adequate of all bio-mathematical approaches since it takes so many sound biological principles into account. The frame for all conclusions is that populations are self-governing systems which regulate their densities in relation to their own properties and to those of the environment. This is done by imparing essential things of the environment below the threshold of favourability, or by the increase or decrease of natural enemies.

The density always depends on intraspecific competition either for a limited niche or by their enemies. These factors hold populations in balance. The very ambiguous word "balance" is defined as: "sustained and effective compensatory reaction which maintains populations in being in spite of eventually violent changes in the environment, and which adjusts their densities in general conformity with prevailing conditions. Far from being a stationary state, balance is commonly a state of oscillation about the level of "equilibrium density" which is forever changing with environmental conditions."

These are very unusual interpretations of the terms "balance" and "equilibrium", but they are the only possible ones if we wish to continue their use in biology. They involve that – except in very few situations – animal populations are always in equilibrium as long as they are subjected to the interplay of governing regulating and controlling factors. These terms are, hence, not identical with the full environmental capacity or the "carrying" capacity, which clearly results from the logistic curve. We merely prefer to free them from even the "equilibrium density" which NICHOLSON still retains and which in our opinion has nothing to do with the new interpretation of the term.

The destructive factors do not, when continued for a long period, necessarily cause an additive mortality on the total population, because they induce largely a redistribution of mortality; the relaxation of numbers releases much other mortality which previously had been caused by other factors such as overcrowding. These compensations are important for the restitution of a balance after the first effects of the shock are overcome.

The cyclic system of the regulation of populations can, of course, be brought about only by density-dependent factors which modify the population density as well as the environment's favourability. Yet other factors, such as weather and climate, also produce profound effects upon animal density (if they are cyclic, they may also cause cyclic fluctuations).

Interaction between populations leads to various types of oscillations, often obscured by seasonal or other cyclic environmental influences. These interferences may cause irregularities with no relation to environmental fluctuations. The heavy intraspecific competition is the important factor for the natural selection of the genetic groups with advantageous properties.

These are the population dynamics in simple situations which are subject in nature to an endless series of complications. All these can be worked into the mathematical equations, but it is doubtful if this is generally necessary

(1) because there are at every moment in every situation usually only one or a very few master-factors which dominate the situation even in the ever-changing complexity of nature,

(2) because of the collective influence of a great number of factors which enter the equation collectively and which the ecologist may solve by observation and experimentation, if desired.

The general base of all mathematical theorems from PEARL and LOTKA to VOLTERRA and NICHOLSON is "that the chance of an individual obtaining a sufficient quantity of the governing requisite to enable it to produce one offspring varies directly with the density of the requisite at any given moment." And all theories of population regulation are complementary as they represent each different groups of population systems. The importance of the lag-effect seems somewhat exaggerated by NICHOLSON, as it does not refer to population changes but only to the intervals from one reproduction period to the next.

Paragraphs 6 and 7 of the principles will be discussed later. It is noteworthy that this elaborate bio-mathematical theorem is the nearest approach today to a theory of selfregulating populations. Of course it is not a theorem built upon inductive conclusions but an intuitive deduction which might be truth for an adherent of PLATO, who inscribed over the entrance of his school: "No non-mathematician may enter here." Yet PLATO also denied that his a prioristic ideas could be corrected or annulled by observation or experiment. They are immutable truths independent of observation or experiment. The same is true of all bio-mathematical theorems. They all arise from the desire to explain observed facts, which are often only partly true as they are built upon insufficient experience.

This is soon followed by the claim that certain assumptions drawn from partial observation be regarded as the sound and immutable basis of biology. This is conspicuous in the case of the importance attributed to density-dependent factors. True, only density-dependent factors may be relied upon to end an outbreak and reduce the population to lower levels, usually to considerably lower levels, in one way or another – by lack of food, disease, parasitation, etc. But just as soon as this reduction is effected, every proof of regulation proper, i.e. by density-dependent factors, is missing.

The result would then be a system of population cycle in which the upper third or upper half would be regulated, while the other part of the cycle is definitely controlled by intrinsic or extrinsic factors only. The bio-mathematical approach is very useful indeed, therefore, and each improvement is an advance for ecology – it is also efficient in many cases to explain the reduction of high population densities. It is, on the other hand, incapable of providing an explanation for the other parts of the cycle, certainly not for the entire population cycle, particularly so long as such "cycles" are not facts but rather hypothetic constructions. This situation finds a full parallel in population genetics. According to SEWELL WRIGHT &

FORD (see Chapter VI) we find a great uniformity of pattern at every population peak due to excessive, density-dependent mutation pressure. At low mutation (= population) pressure, at the beginning rise of the population we find a very varied multitude of patterns, as any density-dependent mutation (= population) pressure is wanting. Then all the less vital recurrent mutants survive and grow to manifestation, while they are rigidly suppressed by the strong density-dependent pressures of the population peaks.

SOLOMON[17] fully recognizes the wide experience of NICHOLSON "to select and support the premises of his theory and again to relate some of his numerous conclusions to known biological facts, his methods remain primarily deductive and the results cannot be accepted as biological principles until they are shown to correspond to the facts."

Sectors of bio-mathematical theorems can obviously be demonstrated, others not. They are interesting and important symbols of the possible trend of factor interplay under certain conditions, all connected with the surpassing of a certain density threshold. Below this threshold it seems without substance – so far, anyhow – and the closed systems of all bio-mathematical theorems can therefore not be accepted. We should also point out that mathematics are a part of the human mind, not (necessarily) of nature*.

## 5. Fish populations

The best illustrations for density-dependent regulation are to be found in lakes and sea. We have long known that every change of fish density or of fishing degree produces changes in the speed of growth and the size of the various age classes among fish. The conclusion therefore is that – aside from limitations set by the law of the minimum – this space is saturated with life. Hence, life there is always regulated, not controlled. It would therefore be well worth our while to delve into details of the population dynamics of fish, so ably presented by RICKER[18].

Fishery theory attempts to predict the catch from a given number of young fish, if their initial size and natural mortality rates are known. Methods have been developed for computing the effects of different rates of exploitation, different minimum size limits, etc. upon the yield obtained. Much progress has also been made in developing methods of determining the actual magnitudes of the population statistics required to make these calculations.

These contributions, however, comprise only half of the biological information needed to assess the effects of fishing and an optimum level of exploitation. Fishing changes the absolute and relative abundance of mature fish in a stock, and the effect of this upon the

---

* See note c on page 267.

number of recruits in future years has often been considered only in the most general manner. The views encountered range from the assumption of direct proportion between the size of adult stock and the number of recruits, to the proposition that the number of recruits is, for practical purposes, independent of the size of the adult stock. The possibility of a decrease in recruitment at higher stock densities has less often been considered. The scarcity of information on this subject is easily explained – it usually requires many years of continuous observation to establish a relation between the size of stock and the number of recruits which it produces. RICKER attempts to summarize the theoretical and factual information available.

Theory of population regulation. Basic in any stock-recruitment relationship is the fact that a fish population, even when not fished, is limited in size, that is, it is held at some more or less fluctuating level by natural controls. NICHOLSON[14] showed that the level of abundance attained by an animal can be affected by any element of the physical or biological environment, the immediate mechanism of control always involving competition (this word being used in a broad sense to include any factor of mortality the effectiveness of which increases with stock density). This was called density-dependent mortality. More strictly, density-dependent cause of mortality should include those which become more effective as density increases and those which become less so. The former are those providing control of population size (concurrent[17], negative[19]). The opposite terms (inverse, positive) refer to density-dependent factors becoming less effective as density increases.

Variation in reproduction produced by factors independent of density. General. A comparison of density-dependent and density-independent reproduction is desirable in order to find possible means of distinguishing the two by their effects on population abundance, particularly since it has been suggested that some of the apparently periodic variations in animal numbers may reflect random variability alone (PALMGREN, COLE). It seems improbable that random variations produce the oscillations and fluctuations observed in animals, for a number of reasons.

Single-age spawners. The increase and decrease in a population of random density-independent fluctuations must in the long run occasionally become so great that the eventual local extinction of the population is a certainty. In nature this can be avoided by independent variations of the neighbouring populations of the same species, by which new immigrants may build up a new population.

In multiple-age spawners, the population has a tendency to rise only.

(1) "A population of single age spawners would vary widely above and below its initial abundance. Eventually, after a

few thousand generations at most, it would either become extinct or, more likely, be fragmented temporarily into small independent units.
(2) "Populations of multiple-age spawners would increase in abundance indefinitely, though not without ups and downs.
(3) "The average peak-to-peak period for lines of single-age spawners would be 4.
(4) "The average peak-to-peak period for multiple-age spawners would be more than 3 and would be the greater, the greater was the number of ages in the spawning stock; in the example used it would be 5 years.
(5) "For single-age spawners, and probably for multiple-age spawners as well, "cycles" having longer periods than these would tend to be apparent in graphs of abundance, because of conscious or unconscious mental suppression of small peaks and troughs, and regularisation of large ones."

Reproduction in the absence of density-independent variability.

Single-age spawning stocks. In Figs. 25, 1–8 the stock is in equilibrium at the density at which the reproduction curve cuts the 45 degree line. Then the stock produces enough, and only enough progeny to replace its current numbers. In all cases when the slope of the right arm of a reproduction curve lies between $-1$ and $-\infty$, equilibrium at the 45 degree line is not merely indifferent (as in the left arm) but unstable. That is, any deflection from equilibrium, no matter how small, initiates a series of oscillations along the right limb whose amplitude increases until the dome of the curve is reached or surpassed. The latter event usually sends the stock back to the right limb and the cycle begins again. No matter where they begin, all such cycles eventually reach the dome of the curve, and a stable oscillation series is established for which the dome is a convenient starting point.

Multiple-age spawning stocks. Permanent oscillations of a more modified type also occur in this class. The general characters of their cycles are:
(1) Cycles occur when the outer part of the reproduction curve slopes downward, provided that this slope begins at some point above the 45 degree line.
(2) Cycles diminish and disappear when the slope of the outer arm of the reproduction curve lies between 0 and $-1$. They are permanent when the slope is numerically somewhat more than $-1$, the exact critical limit depending upon the amount of mixing of generations in the spawning stock and the interval to first spawning.
(3) Period of oscillation is determined by the mean length of

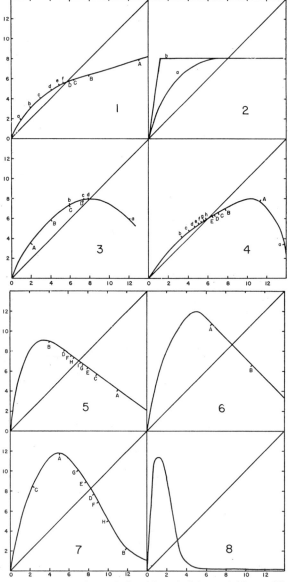

Fig. 25. Stock-reproduction relationships: 1–4 in a stable equilibrium, 5: an oscillation approaching a stable equilibrium, 6–8: in an oscillation. Abscissa: number of eggs produced by parent stock in a given year; ordinate: number of eggs produced by the progeny of this year. The straight diagonale, the 45-degree line, is a useful line of reference: a to h indicate the return to equilibrium; A to I are the extent of the oscillation (From RICKER, W. E., Stock and Recruitment, J. Fish. Res. Bd. Canada, II (5) (1954), pp. 559—623, figs. 1—8).

time from parental egg to filial eggs, being twice that interval or close to it. It is independent of the exact shape of the reproduction curve, and also independent of the number of generations in the spawning stock, provided there is more than one.

(4) Amplitude of oscillation depends partly on the exact shape of the reproductive curve. It tends to decrease with the increase in the number of generations comprising the spawning stock. It increases rapidly with the increase in the number of generations between the parental egg and the first production of filial eggs, up to a limit imposed by the shape of the reproduction curve. When reproduction by a brood begins strongly in the generation following its birth, the oscillations are so weak that they could not be recognised in practice.

Types of reproduction curves. "The average resultant of compensatory and non-compensatory mortality in various environmental conditions is represented by the average size of maturing brood, or recruitment, which each stock density produces. This graph showing the relationship between an existing stock, and the future stock which the existing stock produces, will be called a reproduction curve."

RICKER labels the abscissa as the number of mature eggs produced by the current year's stock, the ordinate as the total of mature eggs produced by the progeny resulting from the current year's reproduction (= sum in period in which the current year's batch is a component of future years' stocks). In nature or with a stable fishery the average size of parental and filial egg production, defined as above, tends to be equal over any long period of years, although striking changes may occur between individual years, or generations.

"Figs. 1–8 show some possible reproduction curves. In each the straight diagonal, the 45 degree line, would describe a stock in which density dependence is absent, the filial generation tending always to be equal to the parental, except as far as density independent factors deflect it. Such a stock would have no mechanism for the regulation of its numbers: if only density-independent causes of mortality exist, the stock can vary without limit, and must eventually by chance decrease to zero. Thus the first qualification of a reproduction curve is that it must cut the 45-degree line at least once – usually only once – and must end below and to the right of it. Figs. 25, 1–8 are taken from actual populations, all having at left an ascending arm to above the 45-degree line, and a dome, in both of which the reproduction is more than adequate to replace the existing stock, and a descending right arm in which reproduction is inadequate to replace existing stock."

All these curves represent the net effect of the sum total of density-dependent mortality factors acting upon the population. The reproduction of any actual year is affected also by density-independent factors, so that the actual number of young produced will deviate from the number indicated by the curve. The first task is to study the isolated effect of the reproduction curve for two situations: first where the currently hatched brood will constitute the whole of the breeding stock, the parents being all of one age-class (most insects, a few small fishes). Second, that two or more age-classes contribute to the spawning stock at any moment (most vertebrates, a few insects, etc.).

RICKER says that no sharp line can be drawn between the kinds of mortality which are compensatory and those which are not, although NICHOLSON, SMITH and others have felt that as a rule biological factors tend to predominate among the former and physical agents among the latter. Among fishes, extremes of water temperature, drought and floods, are physical agents which may often cause mortality whose effectiveness is independent of stock density, whereas deaths from such biological causes as disease, parasitism, malnutrition, and predation will usually become relatively more frequent as stock density increases.

Exceptions to the rule above are sufficiently numerous to make the rule itself of doubtful applicability. For example, the biological factor of predation may have a uniform effectiveness over a considerable range of prey abundance, or at times may even become more effective at lower prey densities; and most, if not all physical causes of mortality are compensatory when stock becomes dense enough that some of its members are forced to live in exposed or unsuitable environments. It is, of course, often difficult to ascribe a death to any single cause.

There is not necessarily any relation between the relative magnitudes of the causes of mortality existing at a given time, as measured by the fraction of the stock which each kills, and their relative contribution to compensation. An important and deadly agent of mortality may be strongly density-dependent, or weakly, or not at all; and different agents may have their maximum compensatory effect over quite different ranges of density.

Age incidence of compensatory mortality. Of the density-dependent causes of mortality, we shall here discuss only those affecting the eggs, larvae, fry and fingerlings. Thus, the relative abundance of a brood will be considered to be determined by the time the first of its females begins to mature – subsequent mortality is assumed to be non-compensatory. This is in contrast to the logistic curve, where the adult population is higher than K, and then a strong compensatory adult mortality would set in. Though such mortality is not impossible, there is in fish populations little indication of it. In any event, the opportunities for compensatory effects are so much greater during the vulnerable early stages that their restriction seems to be a useful approximation.

Fishing is here considered as only attacking mature individuals. Recruitment includes both the commercial-sized fish and the number of maturing fish, both resulting from its reproductive activity. Reproduction here means the number of young surviving to any specified age after compensation is practically complete – it does not include or imply the initial number of eggs or neonate young.

Compensatory kinds of population control.

(1) Prevention of breeding by some members of large populations because all breeding sites are occupied. Territorial behaviour may restrict the number of sites to a number less than what is physically possible.
(2) Limitation of good breeding areas, so that with dense populations more eggs and young are exposed to extremes of environmental conditions, or to predators.
(3) Competition for living space among larvae or fry, so that individuals must live in exposed situations. This too is often aggravated by territoriality – that is, the preemption of a certain amount of space by an individual, sometimes more than is needed to supply necessary food.
(4) Death from starvation or indirectly from debility due to insufficient food, among the younger stages of large broods, because of severe food competition.
(5) Greater losses from predation among large broods because of slower growth caused by greater competition for food. Usually the smaller an animal is, the more vulnerable it is to predators. Hence, slowing up of growth makes for greater population losses. Since abundant year-classes of fishes have often been found to consist of smaller-than-average individuals, this may well be a very compensatory mechanism among fishes.
(6) Cannibalism, i.e. destruction of eggs or young by older individuals of the same species, can operate in the same manner as predation by other species, but it has the additional feature that when eggs or fry are abundant the adults which produced them tend to be abundant also, so that percentage destruction of the initially denser broods of young automatically goes up – provided the predation situation approaches the type in which kills are made at a constant fraction of random encounters (*Tribolium*, muskrats).
(7) Larger broods may be more affected by macroscopic parasites or micro-organisms, because of more frequent opportunity for the parasites to find hosts and complete their life cycle.

(8) In limited aquatic environments there may be a "conditioning" of the medium by accumulation of waste materials that have a depressing effect upon reproduction, increasingly as population size increases. (Note also the contrary effect, as explored by ALLEE).

These compensatory effects are likely to be complementary to a considerable extent. If, for example, exceptionally favourable conditions permitted a good hatch of even a large spawning of eggs, a reduced growth rate of the fry would permit increased predation and so reduce survival.

Combinations of compensatory and non-compensatory mortality.

"In natural populations non-compensatory mortality is superimposed upon the reproduction expected from density-dependent factors, and the curve of population fluctuation is the result of both. The random effect can be introduced either before or after the amount of reproduction indicated by the curve is written down, according as the random factors are thought to act before or after the compensatory ones. The latter procedure gives greater influence to the random element, and may correspond better to events in nature, though the two types of mortality may of course act concurrently, and to a considerable extent probably do."

Remarks: Environmental physical factors act by no means so arbitrarily as is usually assumed (See the seven fat and lean years of the Bible, Wellington, etc.). In nature physical mortality seems usually to occur before any compensatory mortality can set in.

"When the random element is given still greater relative importance, the reproduction-curve element becomes unidentifiable as such. However the latter continues to make an important contribution to the resulting population changes. It makes peaks and troughs much less numerous than they would otherwise be and it provides a control of limits of abundance – that is, the progressive upward tendency (in the multiple age spawners) if the random curve is effectively curbed. The relative importance of random and compensatory factors in determining population abundance also depends somewhat upon the number of ages represented in the spawning stock. When this is large, random factors are less effective in disturbing reproduction-curve cycles."

Effects of removal of mature stock. If man wishes to reduce a single-age population, e.g. in an insect pest, to as low a level as possible by increasing destruction, he can be sure of making direct progress towards this goal only if the reproduction-curve is like those of Fig. 25,1 or 2. If it is like Fig. 25,3 or 4, and stock happened to be on the outer arm, some moderate rate of removal would dampen a crash which would otherwise be imminent. With curves like 25,5 to 8 a paradoxical situation develops. Moderate destruction of adults will in general tend to stabilize the population at or about some magnitude greater than its primitive average abundance. In Fig. 25,7 maximum abundance is consistently achieved when as much as 60 % of the spawning stock is destroyed each year. To reduce the population below the primitive average, more than 73 % must be destroyed. Thus, although sufficiently intensive effort will be successful, moderate destruction of the mature population is worse than no action at all. Furthermore, if an intensive campaign of continuing control of a variable species is decided upon, it is most efficient to begin it at a time that the population is at a low point of its cycle, i.e. when the pest may be doing no particular damage.

The fishery effects upon multiple age group stocks on populations which undergo regular oscillations in the absence of fishery are threefold:
(1) Their average equilibrium abundance is increased at first, but later decreases if exploitation becomes sufficiently intensive;
(2) The amplitude of oscillation decreases;
(3) The period of oscillation tends to decrease slightly.

Some reproductive cycles. Five reproductive cycles of fishes,

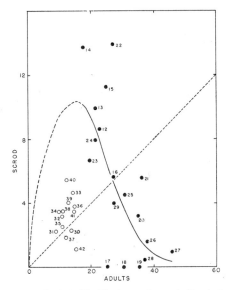

Fig. 26. Abundance of adult Haddock on George's Bank during the spawning season (February to April), related to the abundance of scrod (age III, mostly) at the same season three years later. Black circles: 1912 to 1929, open circles: 1930 to 1943 (From RICKER, W. E., Stock and Recruitment, J. Fish. Res. Bd. Canada, 5 (II) (1954), pp. 559—623, fig. 24).

and those of *Drosophila, Daphnia, Asterias* and *Callosobruchus*, agree fairly well with the preceding theoretical considerations.

Some details of the Haddock *(Melanogrammus aeglifinus)* of the George's Bank population[20] call for discussion. This analysis is of special value because the relevant market reports include an estimate of the smallest commercial size (scrod = mostly age III) independently and in addition to the larger sizes. This yields an estimate of recruitment from year to year, which in turn reflects success of reproduction. When the size of the spawning stock in 1912 to 1929 is plotted against scrod produced (Fig. 26 black circles) a steep outer arm of a recruitment curve is apparent. The 45 degree line has been located very approximately by considering that a pound of scrod during those years had an expectation of egg production, throughout its life, equivalent to the actual egg production in one year of 10 pounds of mature haddock; and by assuming that scrod are half as vulnerable as adult fish to the trawlers. The period studied covers two complete oscillations. The through-to-through periods were 9 and 8 years, and the through-peak amplitude was about 1 : 2. The observed period implies a median age of 4 or 5 years for the production of eggs by the fish of each brood, which is consistent with what is known on the age structure of haddock populations.

## 6. *Urophora jaceana* and *Pieris rapae*

Our next case is that of a terrestric insect, the natural control of the fly *Urophora jaceana*, developing in the inflorescenses of the weed *Centaurea nemoralis*, studied by VARLEY[21], a very special case in more than one direction. The weed is common, and the fly is restricted to it. Its egg is laid within the flowerhead of *Centaurea*, where it is physically protected except against the mice devouring the flowerheads, against feeding by the larvae of phytophagous caterpillars living in the same flower, interspecific competition of the young maggots and, of course, against parasite effects. But apart from *Eurytoma curta*, all other parasites or predators, such as the Chalcidid *Habrocytus trypetae* or the Cecidomyid *Lestodiplosis miki*, are not specific, have many hosts or preys, and are therefore not strictly density-dependent. This latter case is very similar to the abundance of predators affecting the populations of *Lachnus pini* and *Matsucoccus josephi* in the pine forests of Northern Israel[22].

The mortality factors of all stages are discussed in table XVIII. The maggot mortality at the beginning of gall-formation was largely due to parasites *(Eurytoma, Habrocytus*, etc.). Later mortality was largely non-specific, killing these parasites and the remaining *Urophora*-maggots indiscriminately. These later factors killed, according to the table, together about 96 % of the maggots in 1935/36.

**Table XVIII.**

The effect of successive mortality factors on the numbers of the knapweed gall-fly *(Urophora jaceana)* found per sq.m. at Madingley

|  |  | % killed | No. killed per sq. m. | No. alive per sq.m. |
|---|---|---|---|---|
| **1934** | | | | |
| VII | No. of larvae in gall-cells | – | – | 43 larv. |
| | Died due to unknown causes | 5 | 2 | 41 larv. |
| | Parasitized by *Eurytoma curta* | 15 | 6 | 35 larv. |
| VIII | Miscellaneous parasites | 39 | 9 | – larv. |
| | Destroyed by caterpillars | | 4.6 | 21.4 larv. |
| **1935** | | | | |
| Winter | Winter disappearance not estimated | ? | ? | 21.4 larv. |
| | Destroyed by mice | 18.5 | 4 | 17.4 larv. |
| V/VI | Miscellaneous parasitism | 60 | 10.5 | 6.9 flies |
| VII | 6.9 flies emerged per sq. m., 42.5% were females. Mean no. of eggs laid: 70 per female | – | – | 203 eggs |
| | Infertile eggs | 9 | 18.3 | 184.7 eggs |
| | Larvae died before forming galls | 20 | 37.1 | 147.6 larv. |
| | Larvae died in galls due to unknown cause | 2 | 3 | 144.6 larv. |
| | Parasitized successfully by *Eurytoma curta* | 45.5 | 65.8 | 78.8 larv. |
| VIII/IX | Miscellaneous parasitism | 37 | 14.0 | |
| | Destroyed by caterpillars | | 14.8 | 50 larv. |
| **1936** | | | | |
| Winter | Winter disappearance | 61.5 | 30.8 | 19.2 larv. |
| | Killed by mice | 64 | 12.2 | 7.0 larv. |
| | Larvae died (by unknown causes) | 26 | 1.8 | 5.2 larv. |
| V/VII | Birds | | 0.4 | |
| | Various parasites | 31 | 1.2 | 3.6 larv. |
| | Drowned in floods | 44 | 1.6 | 2.0 flies |
| VII | 2.03 flies emerged per sq.m. 42.5% were females. Mean number of eggs laid: 52 per female | – | – | 44.8 eggs |
| | Infertile eggs | 15.3 | 6.9 | 37.9 eggs |
| | Larvae died before gall formation | 26.2 | 9.9 | 28.0 larv. |
| | Larvae died in galls by unknown cause | 4.3 | 1.2 | 26.8 larv. |
| | Parasitized by *Eurytoma curta* | 27 | 7.2 | 19.6 larv. |
| VIII/IX | Miscellaneous parasitism | 36 | 3.7 | |
| | Killed by caterpillars | | 3.3 | 12.6 larv. |

The main density-dependent factors acting upon the population density were:

(1) Early maggot competition within the flowerheads reduced the number, but if no other decimating factors would have acted, would still have produced a density of **2400** flies per sq.m.

(2) The fecundity of *Eurytoma curta* was reduced by lower populations of *Urophora*. The area of discovery was calculated according to the theory of finding at random search. VARLEY correctly assumes that the at random search in the basic equations of NICHOLSON includes the restriction to a search for the particular host, meaning only that the rate of discovery of new hosts at any instant is "proportional to the product of the population densities of the searching parasites and the undiscovered hosts." The estimated area of discovery was 0.25 sq.m., the steady density of *Urophora* which it could maintain as the only factor operating after gall-formation was calculated to be 0.68 adult *Urophora* per sq.m., a cipher well below the observed figures.

(3) The non-specific parasite *Habrocytus trypetae* has two or three annual generations – as compared to the one annual generation of *Urophora*. Its adults emerge at seasons when few or no suitable stages of *Urophora* were available. This brought about a great fall in its population level and prevented it from being an effective regulating factor.

The effect of all non-specific mortality, which kills equal fractions of *Urophora* and of *Eurytoma*, is a reduction of the importance of this parasite as a regulating factor. Thus the population level of the host is raised, just as the third law of VOLTERRA[5] or BODENHEIMER & SCHIFFER[23] assume. Combining all mortality factors, NICHOLSON's theory "gives calculated steady densities of host and parasite which are in good agreement with the observed facts."

VARLEY justly refutes the conclusion proposed in our "Problems" that all factors are of destructive value in direct proportion to the percentage per stage destroyed by each. The last decades have taught us the principle of compensation, according to which the raised mortality of any one factor does not necessarily induce a higher total mortality. This has already been discussed in connection with ERRINGTON's work. The influence of differential mortality by various factors is brought out by the following table of VARLEY (Table XIX).

A careful study of VARLEY's paper is recommended, despite its limitations, mainly caused by the fact that it is the result of two years of observation whereas some decades are required in order to come to valid conclusions. It may serve as a symbol of what can be achieved by mathematical interpretation of observed facts, once density-dependent regulating factors are in play*.

Not always is the host limited in numbers. It is astonishing, at first view, that the introduction of the common European White *Pieris rapae* into parts of North America led to the extermination of its two congeners *Pieris protodice* and *P. oleracea* over wide areas[24]. Now their hosts are

---

* See note d on page 267.

**Table XIX.**

The theoretical effect of different types of mortality factors on the balance between *Urophora* and *Eurytoma curta*
Natural rate of increase of *Urophora*: 18, area of discovery of *Eurytoma*: 0.25 sq.m.

|  | Steady density of the gall-fly | | Steady density of adult *Eurytoma curta* | Percentage of gall-fly larvae parasitized |
|---|---|---|---|---|
|  | Adults per sq.m. | Available larvae | | |
| A. *Eurytoma curta* acting alone | 0.68 | 12.2 | 11.5 | 94.5 |
| Mortality acting only on gall-flies: | | | | |
| B. Before *Eurytoma* attacks | | | | |
| (1) 50% mortality | 1.1 | 10 | 8.8 | 89 |
| (2) 90% mortality | 2.9 | 5.3 | 2.4 | 45 |
| (3) 92% mortality | 3.3 | 4.8 | 1.5 | 31 |
| C. After *Eurytoma* attacks | | | | |
| (1) 50% mortality | 0.55 | 10 | 8.8 | 89 |
| (2) 90% mortality | 0.29 | 5.3 | 2.4 | 45 |
| (3) 92% mortality | 0.26 | 4.8 | 1.5 | 31 |
| D. Mortality acting only on *Eurytoma*: | | | | |
| (1) 50% mortality | 1.35 | 24.4 | 11.5 | 94.5 |
| (2) 90% mortality | 6.8 | 122 | 11.5 | 94.5 |
| (3) 92% mortality | 8.5 | 153 | 11.5 | 94.5 |
| E. Non-specific mortality acting on gall-flies and *Eurytoma* after its attack: | | | | |
| (1) 50% mortality | 1.1 | 20 | 8.8 | 89 |
| (2) 90% mortality | 2.9 | 53 | 2.4 | 45 |
| (3) 92% mortality | 3.3 | 60 | 1.5 | 31 |
| (4) Over 94.5% mortality | 0 | 0 | 0 | – |
| F. Census data for comparison: | | | | |
| 1935 | 6.9 | 147 | 1.9 | 42 |
| 1936 | 2 | 28 | 1.5 | 27 |

wild cruciferous herbs which abound and never show any sign of decimation by the Whites, except very occasionally and very locally. Competition of the caterpillars for food or for other niches of development cannot be cause. *Pieris rapae* was apparently introduced without its native parasites, and those of the native Whites did pass over to it only after an initial period of hesitation. This same phenomenon has been observed by us in Israel on the occasion of the initially devastating invasion of a citrus area by the so-called *Pseudococcus comstocki*. Only after about two years was this species heavily attacked by the native parasites and predators of *Pseudococcus citri*. Both species were thereafter reduced to very low levels from which they have not yet recovered, after fifteen years.

During this parasite-free interval the very vital (euryoecous) *P. rapae* had the opportunity to increase favourably in numbers. Let us assume that the native Whites before the invasion of *P. rapae* had a steady host/parasite density of $n$. As soon as *P. rapae* was also attacked by the same parasites the host population changed abruptly to a considerable rise. A rapid density-dependent increase of the parasites set in, until the level of the entire host population was below its earlier level, subsequently returning to that connected with the host/parasite density $n$. Now the most abundant species of Whites was the euryoecous *P. rapae*, meaning that the absolute numbers of *P. protodice* and *P. oleracea* were much smaller than before, in the new steady state. Every subsequent fluctuation had the same effect – still increasing the density-dependent decimation of the native Whites, leading after a few such fluctuations to their eventual extermination.

This mechanism is effective while eco-niches and especially food plants were abundant, utilised only to a very small degree, whilst other density-dependent factors – apart from food – were a causal agent. This case of *Pieris rapae* serves to illustrate the limited importance of the *Urophora* situation.

## 7. Climate and weather influences on population levels

In turning to the single groups of factors in play for the determination of a population level, we begin with the discussion of the climatic and weather factors and of the processes which they initiate.

It is generally agreed that climate is a very important determinant for the numbers of a species. Often these statements are qualified – LACK[25] says that even such climatic mortalities may actually be caused by starvation and similar factors, which may be density-dependent. NICHOLSON[15] agrees in principle that physical factors are important causes of destruction to a lower or greater degree, i.e. determining the variation of the population, but they cannot possibly regulate populations. In his mathematical biosystem this is of course a banality, once the terminology is agreed upon. If all conditions were favourable for a population increase, it could increase indefinitely, if no other factors interfere by changing conditions of favourability, which the enormous density increase would bring about *eo ipso*.

These regulatory influences are not the weather itself but governing mechanisms. Density-reactions in the population itself determine the level or equilibrium of a population finally. Environmental influences, such as seasons, may induce shorter endogenous cycles to become longer in order to step in with the seasonal rhythm; the intensity of endogenous oscillation may also be increased or decreased considerably by environmental influences.

The conclusion is: "In spite of their inability to govern population densities, climatic factors have an influence of outstanding importance upon populations, but their influence is purely legislative. This influence can be represented in the basic equations by appropriate changes in the values of one or more of the primary parameters" (of the population equations).

There is general agreement on the conclusion that climatic or weather factors have no regulative importance, once we have accepted the density-dependent meaning of the term "regulation". This thesis has even one experience in its favour, namely, that most animals are easily and speedily replacing any catastrophic weather decimation within their normal range of distribution.

The negation of climate as a determining influence, however, is supported by the assumption[26] that the weather sequence of the different years is entirely at random, i.e. arbitrary and at chance. Recent experience tends to indicate the contrary: – the number of cases is rapidly increasing where cyclic types of weather changes, respectively a massed alternation of "fat and meagre" years, have become established. This is not only true for area averages, but also for the local weather during short periods, such as April, as a month often determining fertility, or June eventually determining the main larval mortality, each month being concerned with other master factors, such as have recently been analyzed in the bifactorial maps of SHELFORD. Accumulative series of cool, warm, dry or wet weather become, in the important months, of increasing significance (Wellington), and this in such a way that in Canada long-term prognostics of increase and decrease of certain forest insects are well under study, based entirely upon the regularities of recurrence and alternation of weather-front and cyclone movements at the proper season. While weather in itself is density-independent and not regulatory as long as chance sequences occur, the oft-observed local more or less regular sequences change the role and the importance of weather influence. Such sequences make the strict application of regulating factors versus decimating or limiting factors to a mere play with words. We have shown that in cyclic or regular weather trends they have a decisive influence on the host: parasite ratio[23].

The relative increase of a parasite population to that of its host is assumed by THOMPSON[27] to follow a potential curve, which should necessarily end with the extermination first of the host, then of the

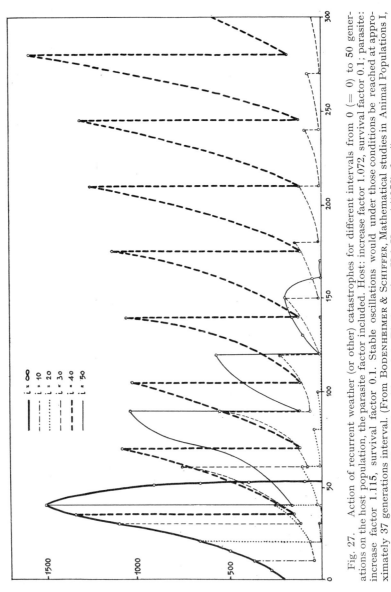

Fig. 27. Action of recurrent weather (or other) catastrophes for different intervals from 0 (= 0) to 50 generations on the host population, the parasite factor included. Host: increase factor 1.072, survival factor 0.1; parasite: increase factor 1.115, survival factor 0.1. Stable oscillations would under those conditions be reached at approximately 37 generations interval. (From BODENHEIMER & SCHIFFER, Mathematical studies in Animal Populations I, A mathematical Study of Insect Parasitism. Acta Biotheoretica, Vol. X no. 4 (1952), fig. 4).

parasite. This conclusion, of course, is unreal and is not observed in nature, where almost all hosts and parasites should long since have disappeared. Actually both exist, fluctuating or oscillating. No accumulation of parasitism from generation to generation is observed in established populations. This would induce us to assume that the parasite is always more sensitive and vulnerable than its host. Yet it is not reasonable to think that this higher sensitivity may be mainly physiological. Let us assume that both are equally sensitive physiologically towards environmental catastrophes. We must consider that the parasite's chain of development is often more complicated than that of the host, which makes the parasite naturally more vulnerable than the host.

BODENHEIMER & SCHIFFER[23] have assumed that environmental catastrophes may directly kill the same percentage of hosts and parasites living. While the same percentage of the parasites is killed as that of the hosts by the direct climatic influence of the catastrophe, let us assume that 50%, half of the parasites which are not killed, survive in dead hosts, all unsuited for further development, and a high percentage of parasite eggs or larvae are killed before they can reach another suitable host. This is just another case of higher ecological sensitivity on the part of the parasite*.

If a climatic or other environmental catastrophe occurs every few years, the relative percentage of parasitation is considerably reduced. Fig. 27 shows this for a few population values. This is a clear case of climatic and other environmental catastrophes changing the effect of population density definitely. With reasonable intervals between catastrophes, such as generally occur in nature without being cyclic or identical, a system of mutual oscillations is introduced by density-independent factors.

It is reasonable to introduce into any formula of population movements a density-factor, certainly effective in dense populations, while uneffective at low ones. Even omitting such a density-factor we may assume that the host population enters a series of years very favourable to its development. The host will rapidly increase to many times the original population, the parasite being of little influence in this initial stage. Suddenly a surprising turn is made. The greater vulnerability disappears, rapidly bringing about the decrease and extermination of the host, before the hyperparasites have the opportunity to influence the parasite numbers.

The following table explains this situation. The numbers are self-explanatory, but the mortality tax should be decreased from 0.88 to 0.44 in better conditions and the survival factor be 0.75 (Fig. 27).

This simple model describes fairly well the major characters of a gradation, as it often occurs: the initial rapid increase of the host as long as the braking parasite effect is not yet felt, followed by the rapid increase if the parasites, the slower increase of the hyperparasites, too slow to inhibit the final effect of the parasites. In reality, these effects usually cease before the actual complete extermination of all populations concerned.

---

* See note e on page 268.

**Table XX.**

Typical gradation of host, parasite and superparasite

|  | Host | Parasite | Hyperparasite |
|---|---|---|---|
| Generation 1 | 10,000 | 100 | 1 |
| Generation 2 | 49,010 | 990 | 10 |
| Generation 3 | 235,240 | 9,810 | 90 |
| Generation 4 | 1,078,955 | 97,245 | 855 |
| Generation 5 | 4,430,448 | 964,327 | 8,123 |
| Generation 6 | 12,586,139 | 9,566,101 | 77,169 |
| Generation 7 | — | 94,927,904 | 733,106 |
| Generation 8 | — | — | — |

There are many more illustrations of the influence of weather and climate on density-dependent processes. If, as NICHOLSON stated, in 1956 at Canberra, a certain niche gives protection against inclement weather and the number of these niches is limited, then at a high population level only a limited number of individuals find refuge in the protecting niches, while the large majority is exposed to higher mortality due to unfavourable weather. It would be decidedly worthwhile to pay more attention to such influences of weather on density-dependent processes.

Another misconception is due to SMITH[26]. He says, even if, for example, 98% of the offspring were killed by climatic factors and 99% destruction was needed for the maintenance of the same number of individuals within the next generation, there would be a permanent increase of population, provided this 1% of difference is not destroyed by enemies. He pretends that climatic factors are for this reason unable to regulate population density or abundance whereas the small percentage killed by enemies does so. That this interpretation gives the true meaning of this school is made clear

**Table XXI.**

Environmental resistance of 32 generations of Red Scale on Palestine

| Generation | I | II | III | IV | Average |
|---|---|---|---|---|---|
| No of eggs per female | 100 | 150 | 100 | 50 | .. |
| Resistance (R ), % | 98.00 | 98.67 | 98.00 | 96.00 | .. |
| 1929 . . . . . . . . | 99.05 | 79.12 | 98.18 | 99.41 | 93.94 |
| 1930 . . . . . . . . | 93.90 | 98.14 | 98.38 | 99.50 | 97.48 |
| 1931 . . . . . . . . | 96.00 | 97.71 | 97.81 | 98.79 | 97.58 |
| 1932 . . . . . . . . | 97.00 | 95.05 | 99.19 | 99.41 | 97.66 |
| 1933 . . . . . . . . | 97.00 | 98.89 | 92.40 | 97.58 | 96.47 |
| 1934 . . . . . . . . | 99.38 | 94.68 | 99.13 | 92.00 | 98.30 |
| 1935 . . . . . . . . | 92.14 | 91.17 | 98.16 | 99.88 | 95.34 |
| 1936 . . . . . . . . | 98.45 | 93.62 | 98.62 | .. | .. |

e.g. by SMITH's figure 129. The fallacy is obvious: it is not correct to assume that climatic destruction – important as it may be – is constant, whereas destruction by enemies fluctuates with population density. Climate fluctuates enormously, and its destructive effect on any species changes considerably from generation to generation. This may be illustrated by total resistance of environment of 32 generations of red scale in Palestine, where the enemy component is almost negligible (Table XXI).

In the following paragraph we give a good number of examples illustrating the influence of parasites, food, and other density-dependent factors. These would be misleading, should the conclusion be drawn that biotic factors always play a highly important role[25]. The author can quote quite a few cases of mass outbreaks from his own limited experience where the share of enemies in the crisis was negligible.

In the spring of 1926 an outbreak of the Vine Hawk Moth *Chaerocampa celerio* occurred throughout the vine-growing districts of Israel. The following generation disappeared almost totally from the vineyards, whereas in our breedings only 15% of the eggs of the 1st generation and 5% of the 2nd generation hatched. No diseases were observed, and parasites as well as predators have certainly remained below 2–3% on the general average.

The almond sawfly *Cimbex humeralis* was in gradation for a series of years at Hartov (1923–6). The larvae were little attacked by predators, and the maximum average parasitism during the peak of the outbreak was 10–15%.

The last two outbreaks of the desert locust *Schistocerca gregaria* in the Middle East cycle broke down due to the delay in prevailing winds at the normal season for remigration to the Sudan and also to drought in the oviposition season.

A discussion of locust populations in general also shows that, although locusts are a staple food of natural enemies in years of outbreaks, the latter's share in breaking a gradation is small and unreliable. Epidemics which may break out only during gradations are entirely dependent upon favourable climatic conditions.

Again, in citrus-scales, where almost annually one or the other coccid is in gradation in groves of a certain age towards autumn and winter, the biotic component in the crisis is almost negligible. If this is true for the red scale *(Aonidiella aurantii)*, it is certainly correct for the mussel scale *(Lepidosaphes beckii)* where even the small amount of parasitism observed in the red scale is absent.

The role parasites play may be approached from still another angle. If parasitism is the regulating factor in animal abundance, superparasitism should logically be the primary controlling factor in parasites. Superparasitism should either be effective in reducing and retarding the effect of primary parasitism in a very efficient

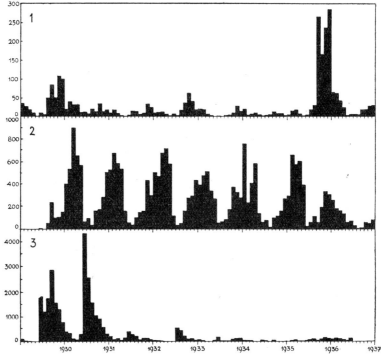

Fig. 28. Monthly fluctuations of three Citrus scale populations in Israel: Red Scale (above), Chaff Scale (middle) and Florida Wax Scale (Below: From BODEN-HEIMER, 1951).

manner or, if not, it puts into doubt the regulatory value of parasitism itself.

A few actual cases may be reported.

| Year | Parasites | | Hyperparasites | |
|---|---|---|---|---|
| | Host cells present | Average % of parasites | Host present | Average % of hyperparasites |
| 1928/9 | 14,638 | 3.4 | .. | .. |
| 1929/30 | 42,386 | 4.7 | 1,078 | 0.9 |
| 1930/1 | 72,772 | 2.6 | 1,764 | 0.4 |
| 1931/2 | 61,879 | 4.2 | 4,381 | 8.6 |
| 1932/3 | 121,238 | 7.2 | 14,202 | 3.4 |
| 1933/4 | 138,878 | 5.7 | 10,560 | 1.6 |
| 1934/5 | 117,679 | 6.1 | 10,131 | 4.0 |
| Average | .. | 5.1 | .. | 3.2 |

Parasites do little damage to the lac crops, and hyperparasites are of little value as controlling agents of parasites of *Laccifer lacca* KERR[28] (See Table page 146 below).

In the open nests of the apple-web moth *Yponomeuta malinella* in Yugoslavia, where parasitism is fairly low, superparasitism on the most important primary parasite *Angitia armillata* is often considerable[29]:

| Parasites no | Nest 1 | Nest 2 | Nest 3 | Nest 4 | Average |
|---|---|---|---|---|---|
| *Yponomeuta* . . . . . | 17 | 22 | 21 | 2 | 16% |
| *Angitia* . . . . . . . | 30 | 39 | 59 | 29 | 37% |

In the case of *Apanteles melanoscelus*, a valuable primary parasite of the Gypsy Moth, secondary parasites seem to be of the greatest importance. The superparasites are in general much less discriminating with regard to host selection than primary parasites. Adult secondaries often puncture holes in primary parasites for the sole purpose of feeding on them, and this is no doubt responsible for a great amount of destruction of primary parasites, combining thus the benefits of parasitism and predation. The 35 species of superparasites indicate that 25 to 30% of the 1st generation's cocoons of *Apanteles* produce adults, while less than 1% of the 2-generation cocoons, carrying the species over winter, produce primary adults the following spring. 50% or more of these cocoons yield neither primaries nor secondaries, as a result of feeding punctures, of strenuous competition, of climate, and of tertiary parasites. However, this case is isolated in its high degree of superparasitism[30].

## 8. Biological effects on population levels

Where is the evidence that parasite fluctuations correspond to those of their host? In the great majority of the parasites with which the author became acquainted during his own experience an abundant food-supply is a conditio sine qua non for their own mass-appearance. Every student of nature well knows that the abundance of parasites in forest-insects, for example, or in scale-insects, aphids, etc. is observed only at the peak of the host's seasonal fluctuation or gradation. The question arises as to whether the differentiation proposed by HOWARD & FISKE that diseases and food-limitation work differently from the parasites because they appear only as a consequence of the host's mass-increase, and therefore cannot prevent the increase, is really justified. Is there any conclusive proof or indication that parasites react in their numerical fluctuations to those of the host which precede its increase? Or does the parasite's increase only follow that of the host after the latter has surpassed a certain threshold, thus limiting it to a certain degree just at or after its population peak, when other environmental resistances, as population density, diseases, etc. and eventually, unfavourable climatic conditions begin to have a reducing effect anyhow? Such is clearly the case in Red Scale and *Chilocorus bipustulatus*, in aphids and Chalcidids, and in voles and their predators in Israel.

There is little doubt that intraspecific population pressure is often a very important point at the peak of the outbreak. Strong competition is intra- as well as interspecifically an important limiting factor, as has been found in Hawaii with regard to parasites of *Ceratitis capitata*[31].

In most animals hitherto studied a reduction in size, fertility, and vitality is induced by this pressure when it has passed the local maximum capacity for the specific stable population. In certain animals, it leads to changes of behaviour ending e.g. in large migrations (locusts, lemmings, etc.). In other animals again, as in many birds, the choice and maintenance of a breeding territory prevents the appearance of any overpopulation a priori by denying reproduction to the surplus[32, 33]. Also, in those animals which live more or less permanently in a saturated environment increase means introduction of unfavourable environmental conditions and lowering of vitality. In such insects as blow-flies and other carrion flies, where there is, as a rule, a surplus of adults in the population, increase of flies does not mean increased offspring, because the limiting factor of suitable carrion present is not changed and maximal oviposition on all carrion is guaranteed, as a rule, by the normal number of flies.

We shall now analyse a few cases of critical years in gradations, i.e. those years which end gradations, and again we choose our

| | No. of individuals | |
|---|---|---|
| | surviving | destroyed |
| No. of eggs to be expected per tree (from 17.4 moths) | 1,200 | |
| No. of eggs actually laid . . . . . . . . . . . . | 250 | 950 |
| Egg mortality (20% Trichogramma + 20% non-hatching eggs) . . . . . . . . . . . . . . . . . . | 150 | 100 |
| Neonate caterpillar (until June 99.3% by weather & constitution) . . . . . . . . . . . . . . . . | 1.100 | 148.9 |
| Old caterpillar (78.9% parasites + 7.7% diverse) . | 0.05 | 1.05 |
| Pupa (16.4 parasites + 14.1 diverse) . . . . . . . | 0.032 | 0.018 |
| Moths hatched per tree . . . . . . . . . . . . | 0.0043 | 0.297 |

examples from forest insects in central Europe and from horticultural pests in Israel.

For *Panolis flammea* we possess satisfactory observations on the crises of the two last gradations in north Germany. The first gradation ended in the following way[34] [Table page 148 below).

The ratio of this generation to the parent generation is 0.004 : 14.4 or 1 : 7,350. If only climatic factors had been effective 0.9 moths would have hatched and if only enemies, 35. This gives a proportion of 1 : 39 in favour of the climatic and constitutional factors, into which the effect of strong intraspecific population pressure is possibly included.

The crisis in the following gradation occurred as follows[35]:

|  | No. of individuals | |
|---|---|---|
|  | surviving | destroyed |
| No. of pupae per sq.m. | 100(relative) |  |
| Pupal mortality (59% predators + 17% diseases + 1% various) | 33.6 | 66.4 |
| Moth mortality (53% weather + 16% predators) | 13.3 | 20.3 |
| Eggs laid | 792 |  |
| Egg mortality (10.4% weather + 1.4% enemies + 1% unfertilized) | 690.6 | 101.4 |
| Caterpillar first stage (16% weather + 8% hunger + 9% predators) | 490.2 | 200.4 |
| Caterpillar later stages (18+18 + 46 + 70% enemies & diseases) | 50.7 | 439.5 |
| During pupation (99% parasites) | 0.5 | 50.2 |

In this case the reduction was 1 : 200, but the destructive power of enemies was 632 times higher than that of the climate, which alone would have reduced the population to 326 pupae only in the next generation. The influence of climate and intraspecific population pressure on the number of eggs laid is omitted in this calculation.

In the year preceding this crisis the number of pupae per sq.m. in a certain area increased from 30 to 50.9[36]. In this year climatic reduction alone was responsible for a reduction from 522.6 eggs per sq.m. to 130.8, the difference being ascribed to enemies.

One such gradation of *Dendrolimus pini* ended[37] (Table page 150 above).

Analysing this development we find that each of the three major factors, if working alone, would have reduced 444 hibernating caterpillars until hatching of the moth as follows:

        Enemies and diseases to 184 individuals
        Lack of food       to 225 individuals
        Climate            to 400 individuals

|  | No. of individuals | |
|---|---|---|
|  | surviving | destroyed |
| Hibernating caterpillars per tree . . . . . . . . . | 444 | .. |
| Hibernating caterpillars (3% parasites + 10% climate + 18% predators and hunger) . . . . . | 41 | 403 |
| Pupal mortality (17% consequence of hunger + 20% parasites + 8% birds) . . . . . . . . . . . . | 23 | 18 |
| Moths (theoretical egg production) . . . . . . . | 1,496 | .. |
| Moths (destruction and prenatal mortality 77%, eggs remain) . . . . . . . . . . . . . . . . . . . | 346 | 1,150 |
| Egg mortality (6% enemies + 8% unfertilized + 36% 'constitution') . . . . . . . . . . . . . . . . | 173 | 173 |
| Young caterpillars mortality (77% climate and constitution + 9% enemies and diseases) . . . . | 24 | 149 |

It is conspicuous that by interrelation and compensation the total effect is much less than that of every isolated factor.

The picture presented during the period from hatching of the moth to hibernation is different when climatic and "constitutional" factors working alone would have reduced the 1,496 eggs to 29 caterpillars. But in the following period, enemies and diseases alone would have killed, until the hatching of the moth, 25.8 out of those 29 caterpillars, climate alone only 2.9.

SCHWERDTFEGER gives a fair summary of the observed facts: "In all stages and ages *Dendrolimus* has died in varying quantity. Destructive factors of all kinds have permanently co-operated sometimes in higher, sometimes in a lower degree on the diminution of the *Dendrolimus* population. It is impossible to say that the calamity has broken down through one cause, e.g. parasites or unfavourable weather conditions. A long sequence of factors, some of which were of greater, others of less importance, have interacted in order to bring about the collapse of the outbreak."

This is true for all analyses given above.

As for the other biotic factors, we will add only brief comments. We have treated food hitherto as a very negligible factor, as we assumed it to be present in abundance. As a rule this is certainly true, as the luxuriant vegetation everywhere, even in the desert, clearly shows. The *Drosophila* bottle has proven that this assumption is not always correct. The emaciated condition of many mammals after winter also demonstrates that if not the abundance of food, at least its availability has a greater importance than previously realized.

The limiting role of diseases is similar. In almost all cases where diseases were believed to be the primary motor for the crisis of outbreaks, a closer inspection revealed that this theory was untenable (e.g. in voles[38, 39]). It is true that the lemmings emigrated into the lowlands often die with symptoms of a bacterial epidemic – but they would have died anyhow, and the epidemic is caused and advanced by unfavourable environmental conditions.

F. M. BURNETT[40] is one of the most brilliant explorers of the ecology of diseases. ZINSSER[40a] had already demonstrated how fundamental changes in the ecology of lice transmitting typhus were produced by the simple fact that man became accustomed to change his shirt at least once a month. Concerning Yellow Fever, BURNETT writes:

"In the jungle, man is only an incidental host for the virus, but when an opportunity arises the virus can become involved in a totally different ecological association: the tropical seaport with its hosts of *Aedes aegypti* breeding in domestic water containers and its crowded human populations. In such a set-up the virus becomes a true human parasite, repeated passage through man being necessary for its survival."

The same ecological background exists between the Tobagan of the Mongolian steppes, with its incidentals of human plague and the human mass-aggregations in the Indian seaports which favoured the mass aggregation, pullulation and vagration of *Rattus rattus*, and brought them into intimate contact with human masses, thus

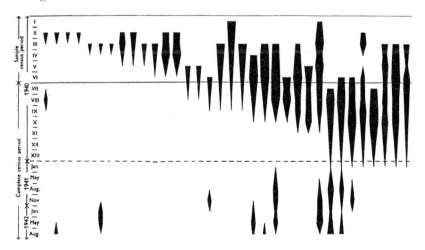

Fig. 29. Population trends of 34 colonies of the desert mouse *Tatera brantsi* infested by a plague epizootic in the northern Orange Free State by width of the vertical columns. Increase, no change and decrease in numbers are represented by widening, parallel and contracting sides repectively. The width of each column bears no relation to the actual size of the population or to the absolute changes in number. Many colonies exterminated by a strong plague infestation in early 1940 were still unpopulated in late 1941 (From DAVIS, D. H. S., Plague in South Africa, J. of Hygiene, 51, (1953), 427—449, fig. 2).

creating secondarily the great centres of human plague. The same conditions have been described by D. DAVIS[41] for the plague of South Africa, the regular plagues of *Tatera* and related mice with incidental human infection. The transition over *Mastomys* to *Aethomys* with a corresponding change of the flea-fauna which become most suited for transition to human beings. Here too the conditions for intensive and extensive human plague outbreaks are given in concentrations of rat populations in human mass-aggregations.

We have quoted similar illustrations earlier, in our section on human malaria, and we could give many other examples where epidemics of animals are dependent upon physical conditions, upon population density, upon change of sensitivity induced by physical factors, by changed habits, etc. In nature epidemics are one of the density-dependent factors which eventually help reduce numbers, but in the absence of which are easily replaced by other density-dependent factors.

LACK[25] recently dealt with the problem of the natural regulation of bird numbers. In his opinion populations are regulated, especially when their number is high, i.e. by density-dependent factors, which also keep their fluctuations within limited numbers. The clutch size of each species is, according to him, a result of natural selection insuring the successful breeding of a maximum number of young, many of which would die early by malnutrition in a high clutch (obviously more by the inability of the parents to feed them properly than by actual absence of food). The breeding season and the number of clutches is determined by the quantity of food. The high mortality of eggs, young and adult birds is described and varies, according to LACK, according to the food supply. This is, however, only to a very limited amount the case in egg-mortality, which fluctuates little in every species, and also in young birds the easy accessibility to prey or exposure is independent of food. He agrees that measurements and observations are inadequate to prove his main theses:

(1) In many cases the other two density-dependent limiting factors, namely, predation and disease, seem to be ineffective;
(2) Birds are more numerous when and where their food is more abundant. It is, however, uncertain if the quantity of food, cover, nesting opportunity, etc. determine this unequal distribution;
(3) Related species are normally supposed to feed upon different food. This is, however, by no means a fixed rule, especially when wintering relatives take the place of the departing summer breeders, and in other cases it may well be induced by micro-habitat, availability of food or other environmental differences of food supply;
(4) Fighting, observed frequently in winter, takes place for food.

LACK thinks the scarcity of starving birds found in nature does not militate against this view. When a bird population is up against the food limit, it seems likely that only a few individuals will be starving at any time. This certainly is a keen assumption.

The bird cycles are explained by him as complex interrelations of predator/prey/food chains, while irruptions, like those of crossbill *(Loxia curvirostra)*, waxwing *(Bombycilla garrulus)* of Pallas' sandgrouse *(Syrrhaptes paradoxus)*, etc. are clearly caused by food shortage in their permanent home. These latter conclusions seem well based and long since generally accepted.

LACK promotes an eventual theory that security of food supply is not the main aim of the territories but only an incidental result of the bird nesting on its feeding grounds, whereas other birds do not feed upon their breeding sites. The benefit of the territories is supposed to be greater safety by dispersion before predation. This is certainly in contrast to our views on the safety of breeding colonies, at least for bigger birds. We have long supported the belief that territorialism is neither the result of securing a quantity of food supply nor of protection against predation, but is produced by a primary behavioural instinct.

While the reproduction rate is little variable and established by natural selection, the main density-control of birds probably comes about through variation in death rate. We have stated before that the facts do not support this theory to such a degree as to make it a general rule.

We are grateful to LACK for his plaidoyer for food supply as the main regulating factor in bird populations. While some facts concerning dispersal and density of bird populations in micro-areas seem to be well established, we must regard the case as a whole to be insufficient to support his main thesis. Exceptional cases such as the increase in the clutches and breedings of certain owls after rodent outbreaks are so far an exception. We seek principally illustrations of a "normal" or "subnormal" population limited by food. As in all plaidoyers of density-dependent types, material is forthcoming regarding peaks of population numbers while material is lacking for populations at their low, and it is precisely there that "regulation" would be most important.

## 9. Biological Control

There would have been no need at all to deal with the control of insects and other animals and plants by biological means, which have been discussed very ably by THOMPSON[42] and SWEETMAN[43], were it not for the reason that even here misunderstandings appear in recent discussions.

H. S. SMITH[26] draws the following conclusion from the author's statements with regard to the predominance of climatic control: "It therefore follows that the introduction of parasites, predators, or diseases from the native habitat of a pest into an area where it has been introduced without them is an unsound procedure." This wholly unjustified conclusion has been drawn neither by the author nor by any other adherent of the school of climatic control. L. O. HOWARD's slogan that all forces must be mobilized for the control of an agricultural, technical, or medical pest is regarded as a basic principle in today's entomology. The fact that the author has published some comprehensive papers on the successful biological control of *Icerya, Pseudococcus* and *Matsucoccus* in Israel should have prevented such conclusions.

The theory really promoted and still maintained is that in climates with distinct seasons the mere introduction and acclimatization of an enemy will be insufficient to obtain the degree of economic control needed, as, for the reasons given above, no progressive accumulation of the parasitism from generation to generation can be expected. Every generation, or every few generations, the parasite must start from the beginning, i.e. from a very low density as compared to that of its host. In order to aid the parasite or the predator it will, as a rule, be indispensable to breed large quantities of the enemy artificially and to distribute them in density centres of the host, year after year. There is nothing new in this conception, as this is the practice which has been in use in California for many years with regard to *Cryptolaemus montrouzieri*. The large-scale "fabrication" of the parasite plays an important role in the literature on biological control. Even in the most efficient cases as in that of *Novius* versus *Icerya* we found it necessary to breed the predator and distribute it to new micro-centres in the citrus-groves of Israel[44].

The only regions where permanent success has been obtained in biological control, after initial introduction and establishment, are tropical oceanic islands, such as Hawaii and Fiji, where climatic conditions are fairly constant and other environmental conditions favourable. Under such conditions accumulative parasitism may be expected. However, even there the effect has often been exaggerated. Two cases, seen by the author personally, may be quoted.

The parasitic Scoliid *Scolia manilae* was introduced against the Oriental beetle *Anomala orientalis*, the grub of which was rather destructive to sugar-cane in Hawaii. Economic damage ceased in the old centres of infestation. The large number of Scolia's flying on every suitable spot throughout the fields of sugar-cane indicates that *Anomala* grubs must still be present in large numbers. Every digging in the soil confirms this. New damage occurs year after year in a narrow zone, mainly in the direction of prevailing winds, which is newly infested with *Anomala*. This damage disappears 1–2 years later, as soon as the *Scolia* has established itself in this zone. The *Scolia* is, therefore, to be desired from an economic point of view, since it has depressed the population density of its host below the level of serious damage. The fact that it still persists in considerable numbers is, however, overlooked.

Reading some reports from Hawaii one might come to the conclusion that the weed *Lantana* has been more or less reduced to insignificance following the introduction of a series of insect pests of that shrub. *Lantana* is, at present, one of the most common plants of Hawaii, covering – especially on lava in the wetter portions of the islands – enormous tracts in compact growth, whereas in the drier parts insects have helped artificial clearing by cultivation to some degree. What really occurred is that extensive cattle-ranging, to which *Lantana* represented a primary danger, has since been more or less abandoned on the islands.

The main reason why no fair judgment as to the value of biological control may yet be given, in most cases, is the fact that no figures at all are available. If a pest decreases following the introduction of a parasite, the importance of the enemy is established. The lack of almost any statistical observations on the mutual fluctuations of both species for a sufficiently extended period and the growing influence of enemies of related species makes it impossible to draw any definite conclusions. The school of biological control should be reproached mainly on this point. And until really competent statistics of this type are available in sufficient number, no final judgment as to the results of enemy-introduction may be given, simply because no scientific analysis is possible. Laboratory experiments and extended breedings of both the prey and the enemy will have to accompany such statistical observations. There is little doubt that TOTHILL and his coworkers, for example, have done a splendid service to Fiji in introducing the parasites of the coconut moth *Levuana viridescens*[45]. But there is little doubt that wind and rain and other, probably climatic, factors inducing "abnormal" mortality amongst young larvae are of decisive influence on the host fluctuation too. Would it not be worthwhile to study such problems from a dynamic and analytical point of view, instead of from one that is partial and empirical only?

May it be suggested that it seems undesirable to extend the conception of the term "biological control" beyond the use of enemies by human agency against pests. Biological equilibrium, biological balance, biotic or natural control would then be reserved for the problem of control of biotic factors under natural conditions.

The economic value of biological control has been firmly established in many cases and is beyond discussion. Unfortunately, the material at hand is insufficient to be treated by the methods of modern population analysis. Such an analysis seems indispensable for the better understanding of real natural phenomena if the theoretical background of biological control is to be thoroughly understood. This is almost as true to-day as it was twenty years ago.

## 10. Biological Equilibrium

The main point in many discussions is the differentiation between density-dependent and density-independent factors. The theory is promoted that only density-dependent factors (as parasites) are able to regulate animal populations, whereas density-independent factors (as weather) may only destroy. This conception was first promoted by HOWARD & FISKE and forms the main topic of the papers published by SMITH and NICHOLSON. This discussion is based essentially on an ambiguous definition of terms.

The reasonable definition of population density seems to refer to the number of individuals present in a unit area. This definition accepted, KIRKPATRICK's[46] conclusion is correct: "The fallacy appears to lie chiefly in the confusion between control of density by balance (for which it is true that "the action of the controlling factor must be governed by the density of the population controlled") and control in the ordinary sense of the word, i.e. reduction of the density, which can be equally well affected by factors, such as climate, which are in no way governed by the density of the population."

We share NICHOLSON's opinion that practically every population lives in an environment filled to capacity, all factors – density-dependent or density-independent – cooperate to manifest any population at the actual level which changes in time and space. Thus, biological equilibrium has assumed in biology a meaning which is very different from that of physical equilibrium: it means simply that under normal conditions every animal species produces and matures just as many individuals as its constitutional potency and the environmental responses permit. The plankton and nekton of the sea or of lakes offer perhaps the best known examples. Yet even in terrestric populations like those of voles, the actual conditions observed in times of the abundance of the vole-cycles, are expressed by rate of growth, body size, etc.[32].

We are still unable to analyse such phenomena for the terrestric fauna as a whole. The new revelations of the theory of compensation open up new vistas which we still do not grasp in significance. In principle, every area is satiated by the populations of all animals present in it. Every population has just as many individuals as it can actually produce and maintain. This by no means excludes ample space in that area for additional species. For a full understanding of all populations we should refer to Liebig's law of the minimum *(sensu latissimo)*.

A basic theory of the "biological school" claims that an insect increases rapidly in number after having been introduced into a new environment, because the native enemies of this insect have not been introduced and the enemy-component is therefore absent or entirely insufficient. We have already shown that this explanation, while occasionally true *(Pieris rapae)* is open to very different interpretations.

Handschin[47] has recently shown that parasites of the buffalo-fly *Lyperosia exigua*, are extremely rare in Java, its native habitat, and had to be bred from other Muscids, before *Lyperosia* could be infested. In Australia, however, where the water buffalo was only recently introduced, Australian parasites are common in *L. exigua*. This shows that the abundance of the fly in Australia is not due to absence of the enemies in the native habitat.

The European corn-borer *(Pyrausta nubilalis)* has recently been introduced into North America, where it has developed to a major pest and has spread with great rapidity over the corn-cultivated area. Its enemies there have hitherto been decidedly negligible (except for the human agency), and the most competent conclusion with regard to its status is[48]:

"It has been demonstrated that corn borer abundance can be correlated with climate and weather, vegetation types and soil types; it can be correlated with soil fertility and cultural practices, such as date of planting, variety, kind of rotation, time of plowing and fertilizer application; and, more significantly still, it can be correlated with the quantity and quality of corn at the period of moth flight."

Again, for the native habitat of *P. nubilalis* the most competent observers state[49] that no definite correlation has been observed "between the quantitative or qualitative composition of the fauna parasitic on the borer and the degree of infestation or extent of damage observed in the various zones. Areas in which parasitism is high are not necessarily those in which Pyrausta is least injurious" and vice versa. The average total parasitism fluctuates in central and southern Europe between 10 and 36%[50]. However, this parasitism works only after a 90% mortality, due to climatic or host resistance, has been passed. The main point seems to have been missed by most students: the change of environment has enormously increased the vitality and the euryoecy of the species in North America.

Thanks to the thorough research made by the American entomologists, we are able to express this change in figures[53]: whereas *P. nubilalis* has been recorded from about 12 hosts in its native habitat, only 1 (corn) being a primary host, the same species has invaded in its new habitat over 200 plants, belonging to 40 different families. Thirty-five plants are definitely recorded as true hosts; for a large number of the remainder this is highly probable, many of them being primary hosts. This sudden polyphagic development is, as has been shown by the author[54], a definite indicator of increased general vigour and greater vitality. The primary reason for the abundance in the new environment is, hence, a physiological reconstruction within the insect stimulated by the new environment, and probably – our knowledge of such phenomena still being at its beginning – decreasing slowly with time.

Another, and perhaps better indicator for the raised vitality of the corn borer in North America as compared with Europe is its fertility: 240 eggs per female are reported from Europe[51] as against 400 eggs per female noted by many American authors[48, 50, 52]. The best explanation refers to the genetic mixture of various strains, comparable to the increasing size of man in recent times due to large population pressure reducing the effect of size-depressing genes in every local population. What occurs is probably often more than a merely passing hybrid vigour after heterosis, but in many cases heterosis may lead to an improvement of standard and vitality, fully explaining the sudden rise of vitality. On the other hand, hybrid vigour must often lead to the breakdown of new populations to lower levels.

A similar explanation may partly illustrate the rapid spread of European sparrows and starlings in North America, but in this case the species introduced were certainly more euryoecous (ecologically plastic) from the start than their native competitors.

One of the species which HOWARD & FISKE[1] had in mind was the Gipsy Moth *(Lymantria dispar)*. It is of the greatest interest to note that KOMAREK[55] has come to the conclusion, after many years of study of the moth in its native and natural habitat, that the climatic resistance of all stages of development of Tachinid flies is lower than that of the gipsy moth. Hymenopterous parasites are still less important and certainly have no regulatory influence at all on the host population.

These few cases, chosen from comprehensive and equally conclusive material, show again that parasites are often a secondary factor in regulating animal populations in their native home and that other factors (probably genetically conditioned), as vitality changes in the organism, are, as a rule, of greater importance than the lack of a few parasites or other enemies.

We fully realize today that changes of vitality occur from population to population of the same species as well as within the same population from year to year. This is another biological factor which cannot yet be included in bio-mathematical equations.

With regard to the regulatory value of predators the conception of HOWARD & FISKE is partly justified. Birds, provided that the local bird fauna is regarded as a whole, prey on animals in the degree of their accessibility[56]. This means that in the end the same proportion of all species is always eaten and that the absolute food quantity is relatively higher when the species is less abundant than when it is at its peak. In years of exceptional abundance of prey the breeding territory of birds is often reduced or, outside the breeding season, more birds than usual are attracted to the district with exceptionally abundant food-supply. The gathering together of birds of prey (which has no reducing importance) in districts of local vole outbreaks illustrates this fact.

Furthermore, there exists a group of catholic predators, some reduviid bugs, carabid beetles, selachians, and cuttlefish, etc. among which the annual or seasonal fluctuations of host abundance exert, as a rule, no influence whatsoever on numerical abundance, for the food present always satisfies their needs. Minor fluctuations may be induced by differential food value of various hosts, but those are mostly superimposed on the major fluctuations dependent on climatic conditions. This is the case in the Indian cotton-bug

**Table XXII.**
Eight years population counts of *Ceroplastes floridensis* and of *Parlatoria pergandei* (on 400 citrus leaves near Hedera)

*Ceroplastes floridensis*

| Year | Month | | | | | | | | | | | | Total | Max. Min. |
|------|-----|----|-----|----|----|-----|-----|-----|-----|----|-----|-----|-------|---------|
|      | I   | II | III | IV | V  | VI  | VII | VIII| IX  | X  | XI  | XII |       |         |
| 1929 | 118 | 39 | 26  | 12 | 9  | 1,891 | 1,298 | 1,783 | 2,248 | 1,568 | 1,365 | 865 | 11,222 | 270 |
| 1930 | 413 | 375| 190 | 119| 325| 4,420 | 2,546 | 1,566 | 1,069 | 941 | 565 | 318 | 12,801 | 37 |
| 1931 | 239 | 83 | 110 | 72 | 165| 495 | 355 | 243 | 74 | 158 | 106 | 93 | 2,195 | 7 |
| 1932 | 55  | 34 | 20  | 14 | 15 | 516 | 433 | 209 | 51 | 80 | 103 | 58 | 1,588 | 37 |
| 1933 | 75  | 46 | 18  | 30 | 58 | 206 | 27 | 37 | 33 | 100 | 105 | 109 | 894 | 11 |
| 1934 | 61  | 42 | 24  | 6  | 8  | 68 | 35 | 24 | 53 | 28 | 28 | 20 | 397 | 11 |
| 1935 | 15  | 30 | 50  | 22 | 18 | 22 | 76 | 93 | 53 | 105 | 177 | 132 | 792 | 12 |
| 1936 | 100 | 125| 104 | 84 | 33 | 149 | 16 | 7 | 16 | 19 | 16 | 15 | 686 | 21 |

*Parlatoria pergandei*

| Year | I | II | III | IV | V | VI | VII | VIII | IX | X | XI | XII | Total | Max. Min. |
|------|---|----|-----|----|---|----|-----|------|----|----|-----|-----|-------|---------|
| 1929 | .. | 121 | .. | 445 | 379 | 52 | 138 | .. | 459 | 90 | 155 | 301 | .. | .. |
| 1930 | 406 | 532 | 891 | 656 | 561 | 87 | 98 | 45 | 158 | 164 | 288 | 506 | 4,392 | 20 |
| 1931 | 523 | 673 | 588 | 537 | 163 | 55 | 44 | 106 | 151 | 158 | 439 | 286 | 3,723 | 15 |
| 1932 | 502 | 475 | 660 | 705 | 584 | 112 | 42 | 59 | 163 | 273 | 267 | 439 | 4,311 | 17 |
| 1933 | 392 | 472 | 516 | 343 | 267 | 57 | 59 | 84 | 182 | 376 | 322 | 267 | 3,337 | 9 |
| 1934 | 757 | 234 | 407 | 580 | 134 | 61 | 84 | 124 | 101 | 120 | 162 | 271 | 2,835 | 32 |
| 1935 | 293 | 669 | 574 | 605 | 398 | 28 | 67 | 112 | 63 | 198 | 342 | 317 | 3,666 | 24 |
| 1936 | 258 | 189 | 167 | 144 | 69 | 73 | 32 | 6 | 15 | 57 | 54 | 88 | 1,152 | 43 |

*Dysdercus cingulatus*, where malvaceous plants abound as food-supply but bug populations are limited by climate[57].

There exist, however, other predators which compete with the most efficient cases of parasites known. The cases of *Novius cardinalis* versus *Icerya purchasi*[44] and of *Symherobius sanctus* versus *Pseudococcus citri*[58] have been analysed by the author in Israel. THOMPSON[59] has reviewed the problem and has demonstrated that there is no basis for the discrimination of predators in favour of parasites, both being probably of equal importance, as good predators are of extremely high destructive value. This agrees fully with our own experiences.

In order to study the question of oscillations around a certain average or equilibrium we must first study the size of those fluctuations. Unfortunately, rather few statistical data, over a sufficiently great number of generations, are at our disposal.

Monthly counts of the Florida Wax Scale *(Ceroplastes floridensis)* in Israel, which are rather representative for insects showing a mediocre range of fluctuation in abundance, over 8 years, i.e. over 24 generations, may be compared with parallel counts of the Chaff Scale *(Parlatoria pergandei)* which represents those insects which show extremely small population fluctuations[60] (Table XXII).

**Table XXIII.**
Gradation of forest insects in Germany (After SCHWERDTFEGER)

| Species | No. of gradations | Accrescence | | | Progression | | | | | Regression | | |
|---|---|---|---|---|---|---|---|---|---|---|---|---|
| | | 1 | 2 | 3 | 1 | 2 | 3 | 4 | 5 | 1 | 2 | 3 |
| *Bupalus pinarius* | 15 | 4.9 | 3.9 | .. | 13.9 | 5.4 | 4.3 | 4.9 | 4.8 | 0.4 | 0.3 | 0.003 |
| *Panolis flammea* | 17 | 6.8 | 6.1 | 3.7 | 15.4 | 12.1 | 1.1 | .. | .. | 0.185 | 0.031 | .. |
| *Dendrolimus pini* | 12 | 5.5 | 2.1 | .. | 5.0 | 7.0 | 3.8 | 4.0 | 2.2 | 0.35 | 0.15 | 0.23 |
| *Sphinx pinastri* | 6 | 4.4 | 1.7 | 2.0 | 2.7 | 2.1 | 1.9 | .. | .. | 0.5 | 0.4 | 0.35 |

| Species | No. of gradations | Decrescence | | Duration of gradation | Product of years of | |
|---|---|---|---|---|---|---|
| | | 1 | 2 | | Increase | Decrease |
| *Bupalus pinarius* | 15 | 0.2 | .. | 11 | 137,478 | 0.00072 |
| *Panolis flammea* | 17 | 0.52 | 0.5 | 10 | 31,470 | 0.0015 |
| *Dendrolimus pini* | 12 | 0.63 | .. | 12 | 13,518 | 0.029 |
| *Sphinx pinastri* | 6 | 0.5 | 0.45 | 11 | 121 | 0.0016 |

Other statistics over long periods are available on voles in Israel and on forest pests in Germany. The latter data have been ably analysed by SCHWERDTFEGER[61]. Series of gradations of four of the major pests are analysed and the gradation divided into different stages. The values in Table XXIII are averages from a series of gradations and show the coefficient of increase in relation to the preceding year.

All these gradations are counted in the winter from the number of hibernating pupae. Large as this increase is, the real fluctuation is still larger. In the interval fluctuations do not cease, but the absolute population values grow extremely small. The "normal" fluctuation in the interval ranges from "nearly zero to 5–10 pupae per 100 sq.m." This means that again in the gradation interval fluctuations are not less than 1 : 100 to 1 : 1,000 and that the increase-factor during the gradation has to be multiplied again by 100 or 1,000 in order to obtain the full amplitude of these pests.

Fluctuations in mammal and bird populations are smaller, but by no means as small as is often supposed. The population cycles of rabbits, lemmings, etc. in subarctic regions are certainly not much below 1 : 1,000, if not higher, and even in larger mammals fluctuations of 1 : 20–50 are common. Fluctuations are not absent in subtropical and tropical climates. The effect of the rinderpest on African wild mammals is well known. In Israel an average fluctuation of 1 : 250 has been observed within the last few years in vole populations, etc.

Discussing these figures with a well-known physicist, the author asked him as to whether in his opinion fluctuations of this amplitude could be considered oscillations around an average stable value. His answer was that in physics no wave exceeding the range of 1 : 10 would be regarded as such. Taking into consideration the complexity of biological processes he could still understand fluctuations within a range up to the limit of 1 : 100 being interpreted in this way. But considering the much larger range manifested sometimes in the natural amplitudes, such conception loses any real meaning and common sense.

There may exist many animal species which show little range of local abundance, as may be expected from the animals of the abyssal or from animals saturating their environment, such as blow-flies, etc. for which the oscillation of physical equilibrium around a state of average equilibrium is a conception to be permitted. But for almost all animals which show large fluctuations within a span of a few generations no such description is admissible.

(1) We were wrong to state that all factors of destruction have a value proportionate to the percentage per stage destroyed by each. Apart from compensating processes, the effect of each factor depends mainly

(a) on the sensitivity of the organism towards the factor (which varies with age),
(b) on the accessibility of the factor towards the species concerned,
(c) on its range of variation and on its dependency or independency. All biotic factors are density-dependent, whereas weather may be independent. The control-complex does not only vary from species to species, but even within the same species from region to region and from season to season.

(2) Most parasites and predators have very little influence on the numerical fluctuations of their host or prey as long as the population density of the latter is low to medium. Only when a certain density is surpassed do they start to increase, and as soon as the regression is advanced (by their action or by that of other factors) they automatically recede to insignificance. The same is true for diseases, population pressure, and food, i.e. for all biotic factors.

(3) Parasites and predators are never indispensable for bringing about the crisis of a gradation. The main biotic factors – enemies, diseases, intraspecific population pressure, and food – will certainly make themselves felt when the gradation has passed the normal environmental capacity. They may co-operate and interchange in varying percentages or one factor may replace the others, depending on the given conditions. This fact remains the true nucleus of the differentiation between density-dependent and density-independent factors. Only the first ones may be relied upon to bring about, in their totality, the crisis of a mass-increase. Climatic regression may bring it about, but need not do so necessarily. This conception includes all major biotic factors, instead of picking out arbitrarily the parasite component, which has no intrinsic preference over the other components (predators, diseases, population pressure, food). Small absolute destruction in later stages by parasites and predators (90% destruction in the egg means 90 killed eggs, 90% destruction in pupae only 9 individuals destroyed!) is therefore of the same importance as higher destruction of the same percentage in early development or even a bigger one as many of the 90 egg-individuals would have been killed, if allowed to develop, by other factors.

(4) The higher sensitivity of parasites, which is observed as a general and basic phenomenon all over the world, is rather complex. It does not necessarily mean that the parasite is always more sensitive to climatic influences, despite the fact that this has been demonstrated in many cases. The parasite is dependent upon a long series of other factors, the more important of which are:
(a) The host must surpass the lower epidemiological threshold of density before any gradation of the parasite may occur, and this high level of density in the host must coincide with favourable climatic conditions for the parasite;
(b) The life-cycle of the parasite must fit into the local seasonal

life-cycle of its host, in order to obtain normal conditions of reproduction. *Exeristes roborator* has been prevented from becoming established in North America, because of the disharmony of its seasonal cycle with that of its host *Pyrausta nubilalis*.

(c) The parasite must fight against the defence system of the prey. This fight may be through active immunization (as is the case, e.g., in *Angitia rapae* in *Pieris rapae*) or through repulsion (as is the case in *Icerya purchasi* living on *Spartium* towards *Novius cardinalis*).

Such complex conditions are one reason why almost no accumulation of parasitism from generation to generation is observed under normal conditions. Yet the greatest change has been brought about by the principle of compensative mortality, which has altered all our earlier concepts of factor importance.

# V
# IS THE ANIMAL COMMUNITY A DYNAMIC OR A DESCRIPTIVE (STATISTIC) CONCEPTION?

## 1. Some definitions of animal ecology

Most contradictions in ecological research results disappear upon quiet analysis. Most of the controversies and misunderstandings are caused by ambiguous definitions, misunderstandings concerning the same term, or translation of philosophical concepts into nature, where they have no place. Hence we find that the intensity and the dogmatism of earlier and present controversies are usually in reverse proportion to the actual amount of adequate knowledge accumulated. To reduce possible misunderstandings we will here introduce a number of definitions, repeat them or present a few new ones serving in the ensuing discussion.

Regulation is any control of numbers which is induced mainly by density-dependent factors, including the effects of compensation. Control is every determination of numbers, irrespective of its induction by density-dependent or -independent factors. This means that every animal population is controlled, but it is regulated only towards its peak of abundancy. This is the best interpretation of the facts as actually grasped.

Every population is in a state of biological equilibrium. We are still poorly informed on the population low. In the present state of our knowledge we cannot possibly apply the cycle-theorem of bio-mathematics to it as far as natural populations are concerned. How far oscillations and/or fluctuations observed in nature observations are homologous or only analogous to those required by the cycle theorems is still altogether unsettled.

It would be well for the peace of the human mind if the bio-mathematical cycle theorems could be accepted. But just as in genetics, the peace of the human mind is not the primary aim of scientific research. The actual fluctuations in numbers are often very different from what is visible to the eye. What changes is often only the bio-mass, as periodically eggs or young are produced abundantly but do not form large masses or catch the eye of the observer. What is actually decisive are the relatively few adults reaching maturity out of this great abundance. The population count must always take into consideration numbers and masses of all stages of development. What NICHOLSON and BAILEY (IV[14]), call lag is not actually a lag of population, but refers to the interval between the mature individuals of each generation.

Terminology in animal psychology has made great strides since von UEXKÜLL (II[35]). The environment, as we see it, is a purely human conception and description. Every species of animals perceives its own "surrounding world", which is a specific sensory and integrational selection of the quasi "objective" environment. There are as many different "surrounding worlds" as live different species in any given area. Frog, fly, predatory bug, butterfly, vole and lion, sparrow and owl see its own very different "surrounding world", each perceived by different sense organs and integrated by different central nervous systems into every animal's specific "surrounding world". We here ignore for our discussion that animal psychology, in its research, calls "worlds" a number of inner tunings which alternatingly dominate the animal behaviour.

We wish to propose the term eco-world as the combination of all exogenous and endogenous factors, processes and organisms which have any relation – whether directly or not directly perceived – to the living species. The species itself is of course – just as in psychology – an integral part of this eco-world. This term is distinguished from all others (except the "ökologische Umwelt" of PEUS[1]) in that it is not an area or spatial concept at all. The species may live in various areas or landscapes, but has always the minimal quantity of its own eco-world at its disposal, without which it cannot exist (by definition).

The eco-system of TANSLEY[2] and SOLOMON[3], the biocoen of German Workers* are all bound to an area. Yet most of this area is ecologically "dead" and insignificant for the species, and hence it is superfluous to include it into the animal's eco-world. It is like the effect of the "farest plane" (VON UEXKÜLL) in animals. Anything moving beyond this farest plane (in flies about one meter's distance) becomes only bigger or smaller as it moves towards or from this plane. Any movement on the part of the animal causes the perception – aside from becoming larger or smaller – to go away or to approach the subject. The action has suddenly assumed a personal tuning to the subject.

Like this farest plane, most space remains without relation or importance to the animal. On the other hand, it is generally not

---

* K. FRIEDRICHS who mottoed each chapter of his 'Ökologie als Wissenschaft von der Natur' (1937) with a sentence taken from the fundamental ecological book of ADOLF HITLER 'Mein Kampf' is the extreme exponent of the German school. We can compare his entire work only with that of a school dead for a hundred years, namely, that of OKEN, KIELMEYER, SCHELLING. Aside from discussions on natural history, the whole contents was devoted in part to abstruse philosophical reasoning, in part to romantic poetry, both of which had been well characterised by HEINRICH HEINE long ago in his treatises on German romanticism. FRIEDRICHS' philosophy has misled an entire generation of German ecologists to sterile discussions on terminology. A few – like THIENEMANN and SCHWERDTFEGER – avoided this trap, as their own wealth of observation was large enough to liberate them partly from the influence of this school. These aberrations do not belong to ecology as a science; actually a reaction has begun from within.

limited to any one area but may appear in certain niches of the most different areas. Thus the word environment, including the animal itself, is used for eco-world. When it is otherwise understood, it is artificial to every species which has its own more limited and more extensive ecoworld.

As niche we regard any micro-ecoworld required for the existence of the species at a certain stage: to the egg, to a larval stage, or to the young, to the feeding, resting, respiring, breeding and successful reproduction of the adult. It is erroneous to expand it to a meaning synonymous with ecoworld, as has sometimes been done.

It is clear that the ecoworlds of many species, like those inhabiting the soil, dense forest, tree crowns, sea-bottom, quiet waters of the open sea or its stormy littoral, sunny open steppes, etc. overlap in part, but the ecoworld of an amoeba, a snail, a fly or a mouse – to mention only certain types of animals – will never be identical, even when the species are much more closely related.

We have so far discussed only terms connected with the ecology of populations of the same species – its autecology. It was CHARLES DARWIN who realized that the sharpest competition takes place among members of the same species. We know of a few great exceptions where the intraspecific competition has been transmuted into mutual aid – in colonies of bees, ants, termites and – let us hope – in future human world-society. Any interspecific struggle is in principle the keener, the more related the species and the more niches they have in common. Close observation of nature also teaches us how rare are the cases where a predator really diminishes the capital or the permanent stock of its prey, not merely its temporary surplus.

The biotope is the smallest area which to the human mind is characterized by certain plants and animals or by a common character of faunal and floral nature – such as an area in the tropical rain forest, where often no common species are conspicuous but where the human observer (the geographer, the ecologist) has not a moment of doubt that the local tropical rain forest is a characteristic landscape of greater or smaller extent, differing despite its lack of statistical coherence in its plants and animals from the surrounding savanna, steppe or high mountain peaks. This is in contrast to the large monotonous expanses of the taiga, the tundra, the bottom or the plankton of the sea, where a few species of plants and animals dominate the geographical area and where the latter may form an essential part of its geographical description.

Thus, the biotope is not an ecological term at all, in that it is more a description of the dominants and predominants of the geographical areas of the human ecoworld. Almost every detailed plant-sociological analysis offers a good illustration of this point of view. The rich vocabulary of descriptive terms of plant-sociology only proves that even in the most monotonous types of vegetation, an essential stability of the prevailing species is combined with an embarrassing instability of almost all other elements. This does not by any means detract the great influence of factors of physical character, such as shade or water, minimal or maximal temperatures, etc. on the faunal and floral composition of any area. The biotope is, as PEUS[1] justly says, a conception of the human surrounding world (Umwelt) and has no significance for the ecological situation of any other animal.

There is no reason to claim that any organisms living in a biotope show higher or lower degrees of integration. We shall later discuss the conception of a **superorganismic biocoenosis** and prove it to be an intuitive hypothesis which may be right or wrong, but is at present in the domain of philosophy and not of science.

Another school (SCHWERDTFEGER[4], SCHWENKE[5]) sees in the collection of all physical, chemical and biological factors of any specific area the primary aim of ecology. It calls the physico-chemical inquiry into the area of its population densities "biocoenotics". We appreciate greatly any penetrating research of this kind, particularly if continued over a long period. This type of research may also contribute to ecological research, but the name is wrongly chosen, as it is primarily of geographical character. Research of this kind – sufficiently prolonged and spatially extended – would permit the elimination of all that is ecologically dead in the area and bring to light the ecoworld, or part of it, of the species concerned. To call biocoenotics (in this sense) the aim of applied zoological research is, however, not in agreement with the true goals of ecological research. It is a potentially useful method, not an aim. Biocoenotics belong in neither sense to ecology, but rather to philosophy or geography.

The true goal of ecology, or at least part of it, is to concentrate on populations whose fluctuations must be analysed not only by single-factor analysis which is often subject to chance result and which is subject to change in importance by subsequent or earlier compensations.

We have given no explanation of the terms **density-dependent and -independent factors**, which seem logically neatly separated. Their epidemiological importance, however, becomes more and more obliterated, as also their identification with climatic or biotic factors. This obliteration occurs principally because

(1) Many density-independent factors also take place in more or less regular rhythm (see e.g. the seasonal hamsin winds, the sequence of fat and lean years, the influence of more or less regular weather catastrophes on the intensity of parasitation, etc.).
(2) Many individuals are exposed to density-independent mortality, as all suitable (= protected) niches are already occupied

by other individuals of the same species – clearly a density-dependent factor.

(3) Epidemics often depend upon certain weather factors, but the spread of the epidemic may eventually depend upon host-density.

## 2. Description of animal communities
## Marine bottom communities

All studies of animal communities must begin with quantitative counts of the animals present in small areas. The analysis of a sufficiently large series of such counts made in a certain region allows a comparison to be made between the species and their numerical abundance in each area and permits the description of the statics of animal communities at the time of the survey.

The study of the environmental factors acting in each area as well as the study of the physical ecology of any species enables us to understand the dynamics of animal communities.

Unfortunately, the number of suitable surveys is rather small. Most of them have been done on the animal communities of the sea bottom. The surveys of PETERSEN[6-8], STEPHEN[9-10], FORBES[11], MOLANDER[12], SHELFORD[13], BERG[14], etc., are outstanding examples of this type of research.

It is, perhaps, most advisable to compile the results of some of these researches in order to obtain a basis for better discussion. PETERSEN[6-8] studied the fauna of the sea bottom in relation to production of fish food in the benthal of the Danish Seas. Quantitative catches with the bottom sampler forced upon him the conception that animal communities are the only means of dividing the bottom fauna into areas and of being able to bonitate (evaluate) the productive value of each area. Beginning with middle-sized communities (= associations of SHELFORD), he succeeded in distinguishing between eight such communities (Fig. 30). Table XXIV gives a fair illustration of his results, which are the average of many catches made throughout the area of each community as well as at various seasons and in different years. His figures illustrating these eight different benthal communities are classical fundamentals of ecological teaching.

MOLANDER's[12] work in the Gullmar Fiord is based on smaller communities (= faciations of SHELFORD) than those chosen by PETERSON. The five communities of PETERSEN represented in this fiord are thus divided into at least 13 (faciations). MOLANDER defines an animal community as "a regularly recurring combination of certain animal types, as a rule strongly represented numerically."

STEPHEN[9, 10], in summarizing the large Scottish survey of the bottom fauna of the North Sea, returns to the conception of animal zones, used in oceanic research by FORBES[11]. A zone is defined as that area of sea-floor of varying width to which certain (characteristic or dominant) species are limited in their distribution. These characteristic species may either show more or less a continuity (e.g. *Tellina tenuis*) or a discontinuity (e.g. *Donax vittatus*) in their distribution. The relative density of the component species within

**Table XXIV.**
The classical communities of marine bottom animals in the Danish Seas as defined by PETERSEN (area unit 0.25 sq.m. of bottom surface)

1. *Macoma* community. On all S. Danish coasts and in the Baltic. 0.3 m. depth at high water, dry at low water.

|  |  | No. | Weight |
|---|---|---|---|
| Lamell.: | *Mya arenaria* | 1 | 26.6 g. |
|  | *Mya arenaria* juv. | 2 | 0.1 |
|  | *Macoma baltica* | 5 | 2.2 |
|  | *Cardium edule* | 1 | 11.6 |
|  | *Cardium edule* juv. | 3 | 0.1 |
| Polych.: | *Arenicola marina* | 4 | 7.3 |
|  | *Aricia armiger* | 1 | 0.1 |
|  | *Nephthys* sp. | 2 | 0.3 |
|  | Total (6) | 19 | 48.3 |

2. *Abra* community *(Echinocardium)*. In the Baltic Sea and the fiords. 16–18 m. depth.

|  |  | No. | Weight |
|---|---|---|---|
| Lamell.: | *Abra alba* | 10 | 1.8 g. |
|  | *Macoma calcarea* | 8 | 2.1 |
|  | *Corbula gibba* | 1 | 0.1 |
|  | *Astarte banksii* | 20 | 12.7 |
|  | *A. borealis* | 4 | .. |
|  | *A. eliptica* | 1 | 0.2 |
|  | *Cardium fasciatum* | 1 | 0.1 |
|  | *Leda penula* | 2 | 0.4 |
|  | *Nucula tenuis* | 2 | 0.1 |
| Polych.: | *Nephthys* sp. | 3 | 2.7 |
| Ophiur.: | *Ophioglypha albida* | 4 | 0.7 |
| Spatang.: | *Echinocardium cordatum* | 1 | 3.6 |
| Crust.: | *Diastylis* sp. | 1 | 0.1 |
|  | Total (13) | 58 | 24.6 |

3. Shallow *Venus* community *(Echinocardium)*. On the open sandy coasts of Kattegat and N. Sea. 10–11 m. depth.

|  |  | No. | Weight |
|---|---|---|---|
| Lamell.: | *Venus gallina* .......... | 18 | 4.4. g. |
|  | *Tellina fabula* .......... | 10 | 0.4 |
|  | *Montacula ferruginosa* .... | 4 | 0.1 |
|  | *Solen pellucidus* ....... | 1 | 0.1 |
|  | *Cyprina islandica* ....... | 1 | 14.4 |
| Gastrop.: | *Philine aperta* .......... | 1 | 0.1 |
| Polych.: | *Travisia forbesi* ........ | 1 | 0.6 |
|  | *Nephthys* sp. .......... | 3 | 0.4 |
|  | *Aricia armiger* ....... | 1 | 0.1 |
| Spatang.: | *Echinocardium cordatum* ... | 3 | 43.8 |
|  | *Echinocardium cordatum juv.* . | 18 | 34.4 |
| Ophiur.: | *Ophioglypha albida* ....... | 1 | 0.1 |
|  | *O. affinis* ............ | 1 |  |
| Crust.: | *Gammaridae* ......... | 7 | 0.1 |
| Actin.: | *Actiniidae* ........... | 1 | 0.1 |
|  | Total (14) .......... | 71 | 99.1 |

4. *Echinocardium-Filiformis* community. At intermediate depths in the Kattegat. 20–22 m. depth.

|  |  | No. | Weight |
|---|---|---|---|
| Lamell.: | *Abra nitida* .......... | 4 | 0.1 g. |
|  | *Corbula gibba* ........ | 1 | 0.1 |
|  | *Cyprina islandica* (3 cm.) ... | 2 | 62.5 |
|  | *Cyprina islandica juv.* (/ 3 cm) . | 1 | 0.1 |
|  | *Axinus flexuosus* ....... | 1 | 0.1 |
|  | *Nucula tenuis* ......... | 1 | 0.1 |
| Gastrop.: | *Aporrhais pes-pelecani* ... | 1 | 2.6 |
|  | *Turritella terebra* ....... | 10 | 4.6 |
|  | *Chaetoderma nitidulum* ... | 1 | 0.1 |
| Vermes: | *Glycera* sp. .......... | 1 | 0.1 |
|  | *Nephthys* sp. .......... | 6 | 2.6 |
|  | *Brada* sp. .......... | 5 | 0.1 |
|  | *Terebellides stromi* ....... | 3 | 0.4 |
|  | *Nemertini* ........... | frag. | 0.1 |
| Ophiur.: | *Amphiura filiformis* ..... | 60 | 7.8 |
|  | *Ophioglypha albida juv.* ..... | 2 | 0.1 |
|  | *O. texturata* ......... | 1 | 0.8 |
| Spatang.: | *Echinocardium cordatum* ... | 5 | 45.0 |
| Crust.: | *Gamaridae* .......... | 2 | 0.1 |
| Pennat.: | *Virgularia mirabilis* .... | 2 | 0.5 |
|  | Total (19) ......... | 104 | 127.9 |

171

5. *Brissopsis-chiajei* community. In the deepest parts of the Kattegat. 60-75 m. depth.

|  |  | No. | Weight |
|---|---|---|---|
| Lamell.: | *Abra nitida* | 2 | 0.2 g. |
|  | *Axinus flexuosus* | 1 | 0.2 |
|  | *Leda penula* | 1 | 1.0 |
|  | *Nucula sulcata* | 2 | 1.0 |
| Vermes: | *Nephthys* sp. | 1 | 1.5 |
|  | *Maldanidae* | .. | 2.0 |
|  | *Balanoglossus kuppferi* | 1 | 2.0 |
| Ophiur.: | *Amphiura chiajei* | 12 | 4.7 |
|  | *Ophioglypha albida* | 2 | 0.7 |
| Spatang.: | *Brissopsis lyrifera* | 5 | 90.0 |
| Crust.: | *Calocaris mandreae* | 2 | 1.7 |
|  | Total (11) | 29 | 105.0 |

6. *Brissopsis-sarsii* community. In deeper parts of the Skagerrak. 186 m. depth.

|  |  | No. | Weight |
|---|---|---|---|
| Lamell.: | *Abra nitida* | 179 | 17.5 g. |
|  | *Cardium minimum* | 11 | 0.2 |
|  | *Axinus flexuosus* | 20 | 0.8 |
|  | *Portlandia lucida* | 1 | 0.1 |
|  | *Leda penula* | 17 | 4.1 |
|  | *L. minuta* | 1 | 0.1 |
|  | *Nucula tenuis* | 14 | 0.6 |
| Polych.: | *Aricia norvegica* | 9 | 6.0 |
|  | *Artacama proboscidea* | 8 | 0.9 |
|  | *Melinna cristata* | 8 | 1.2 |
|  | *Pectinaria auricoma* | 1 | 0.1 |
|  | *Eumenia crassa* | 1 | 1.3 |
|  | *Myriochele heeri* |  | 3.3 |
| Ophiur.: | *Amphiura elegans* | 2 | 0.1 |
|  | *Ophioglypha sarsii* | 5 | 0.4 |
| Spatang.: | *Brissopsis lyrifera* | 9 | 36.9 |
| Crust.: | *Crustacea* | 1 | 0.1 |
|  | Total (17) | 287 | 73.7 |

7. *Amphilepsis-Pecten* community. In the deepest waters of the Skagerrak. 320 m. depth.

| | | No. | Weight |
|---|---|---|---|
| Lamell.: | Abra longicallis | 1 | 0.3 g. |
| | Nearea obesa | 1 | 0.1 |
| | Axinus flexuosus | 16 | 0.8 |
| | Portlandia lucida | 1 | 0.1 |
| | Pecten vitreus | 8 | 0.6 |
| Scaphop.: | Siphonentalis tetragona | 1 | 0.1 |
| Polych.: | Aricia norvegica | 3 | 1.7 |
| | Nephthys sp. | 3 | 0.3 |
| | Terebellides stromi | 1 | 0.1 |
| Ophiur.: | Amphilepis norvegica | 17 | 0.9 |
| | Total (10) | 52 | 0.5 |

8. *Haploops* community. Locally in the SE. Kattegat. 27 m. depth.

| | | No. | Weight |
|---|---|---|---|
| Lamell.: | Venus ovata | 1 | 0.5 g. |
| | Cardium fasciatum | 2 | 0.5 |
| | Axinus flexuosus | 2 | 0.2 |
| | Leda penula | 1 | 0.3 |
| | L. minuta | 1 | 0.1 |
| | Lima loscombii | 3 | 0.8 |
| | Pecten septemradiatus | 2 | 17.9 |
| Vermes: | Aphrodite aculeata | 1 | 10.7 |
| | Clycera sp. | 1 | 0.4 |
| | Eumenia crassa | 3 | 5.9 |
| | Maldanidae | frag. | 0.7 |
| | Pectinaria auricoma | 1 | 0.4 |
| | Terebellidae | 1 | 0.2 |
| | Balanoglossus kuppferi | 1 | 0.2 |
| | Nemertini | 3 | 0.6 |
| Ophiur.: | Ophioglypha albida | 1 | 0.2 |
| | O. robusta | 30 | 1.7 |
| Echin.: | Strongylocentrotus drobakiensis | 1 | 2.9 |
| Crust.: | Haploops tubicola | 875 | 7.5 |
| | Moera loveni | 1 | 0.2 |
| | Verruca stromii | 1 | 0.1 |
| | Total (21) | 932 | 52.0 |

the zone may, however, vary from time to time. The respective areas of distribution of these dominant species may also vary in position and extent.

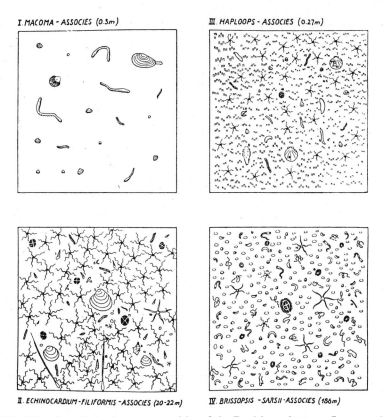

Fig. 30. So-called animal communities of the Danish sea bottom. Compare the proceeding table. The figures in brackets refer to depth see Table XXIV. (After Petersen).

The conception which Stephen[10] presents from the bottom fauna of the North Sea is based mainly on dominant molluscs, echinoderms being of secondary importance in one zone. The zonation is defined only for the areas with sandy and muddy bottom. The areas with a rocky bottom do not only have a different fauna, but apparently also a different major zonation.

### Table XXV.
Major animal zones of the North Sea (after Stephen).

I.  Littoral Zone. From high-water level to 4 m. depth.

   a) On muddy ground with *Macoma baltica*, *Cardium edule*, and *Paludestrina stagnalis* as the dominant species.
   b) On sandy ground with *Tellina tenuis* as the characteristic species.

II. Coastal Zone. From 4 to 40 m. depth in the northern North Sea and 60 m. in the southern North Sea.

    a) On muddy ground with *Abra alba* and *A. nitido* as the dominant species.
    b) On sandy ground with *Tellina fabula* and *Nacula nitida*, or *T. fabula* alone as the dominant species, in Scottish waters. In the southern North Sea, in addition to the above a species *Mactra stultorum* and *Spisula subtruncata* seem also dominants. The Ophiurids *Ophiura texturata* and *O. albida* are confined to this zone.

III. Off-shore Zone. Occurs over most of the northern North Sea outside the coastal zone. The average density of Lamellibranches is low, no species being numerically dominant, but *Dentalium entalis* is widely distributed and may be taken as one of the most characteristic species.

    The Echinoderms occur here in greatest numbers, *Ophiura affinis*, *Amphiura chiajei*, *A. filiformis*, and *Echinocardium flavescens* being the commonest and most widely distributed species.

IV. *Axinus* (= ?) + Foraminifera Zone. In the deep north-eastern portion of the northern North Sea and possibly extending along the edge of the Norwegian Deep. Characterized by the presence of *Axinus* (= *Thyasira*) *flexuosa* and Foraminiferan tests. All species are dwarfed.

Glancing over some of STEPHEN's distribution charts of single species, we must agree with him that the zone is a better definition of conditions in the North Sea than are MOLANDER's communities. The fauna of the coastal zone in the North Sea has a slightly different composition in its northern and in its southern parts, the transition taking place gradually. More "off-shore" species intrude into the northern than into the southern coastal area, and important differences in the dominant species occur. Still greater difficulties arise in the off-shore zone, where a very definite change takes place on passing from north to south. "Species are gradually eliminated one after another, but there is no sharp transition and no natural break which would justify a separation into communities."

The littoral zone with its broken mosaic of varying textures and its many half-enclosed areas is quite favourable to the community conception. The difficulty, however, lies in the fact that each observer describes "new communities which are merely combinations in variable proportions of the same few species characteristic of one or two zones at most. Had the Scottish investigations not extended beyond the littoral zone the community concept would have been quite acceptable, but it becomes less acceptable on pushing the investigations as far as the coastal zone, and when the off-shore zone is reached it breaks down completely."

In studying some marine biotic communities of the Pacific coast of North America SHELFORD and co-workers[13] obtained fundamental results. Their research, again, was limited to littoral communities which fit best into the community concept. Five communities of major rank (= biome) were differentiated with a series of subdivisions (associations and faciations). His type of classification is explained by the following list:

### Table XXVI.
The *Strongylocentrotus-Argobuccinum* biome on the Pacific Shore of North America.
(After SHELFORD).

0–225 m. depth. Salinity and hydrogen-ion concentration fairly high. Plantation of medium abundance. Penetration of light in summer fairly great. Large algae of many species abound. Circulation of water due to tides is the most important climatic factor. Number of animals per 10 sq.m. (minimum to maximum).

A. Dominants

| | |
|---|---|
| *Strongylocentrotus drobachiensis* | 40–200 |
| *Argobuccimun oregonensis* | 2–50 |
| *Balanus nubilis* | 5–50 |
| *B. pugetensis* | 10–400 |
| *Calliostoma costatum* | 2–60 |
| *Psolus chitonoides* | 1–10 |
| *Trichotropis cancellata* | 2–150 |
| *Pecten hericius* | 15–1,000 |
| *Pododesmus macroschisma* | 2–50 |
| *Amphissa columbiana* | 5–85 |
| *Crepidula nivea* | 10–50 |
| *Oalyptraea fastigiata* | 3–6 |
| *Orchasterias columbiana* | 1–2 |

B. Dominants changing with depth

*Strongylocentrotus franciscanus* 0–36 m.: few; 36–125.: 4–60.
*Stichopus californicus:* 5–30 m.; 35 m.: rare.
*Modiolus modiolus* 0–36 m.: absent; deep water: 10–3,000.

C. Influents

*Icelinus borealis*
*Myoxocephalus polyacanthocephalus*
*Rhyamphocottus richardsoni*
*Hyas lyratus* . . . . . . . . . . . . . . 2–10
*Cancer oregonensis* . . . . . . . . . 1–20
*Oregonia gracilis*
*Pandalus danae*

D. Characteristic, but rare species

*Styela stimpsoni*
*Munida quadrispina*
*Crago munita*
*Spirontocaris prionota*
*Hapalogaster mertensii*
*Pagurus kennerlyi*
*Purpura foliata*

I. *Strongylocentrotus-Pugettia* Assiocation

A. Dominants

| | | |
|---|---|---|
| Cucumaria miniata | . . . . . . . | 2–20 |
| C. chronhjelmi | . . . . . . . | 2–30 |
| Crepidula adunca | . . . . . . | 10–300 |
| Puncturella cucullata | . . . . . . | 3–6 |
| Petrolisthes eriomerus | . . . . . | † † † |

C. Characteristic species

*Doriopsis fulva*
*Cryptochiton stelleri;*
*Ischnochiton interstinctus*
*Polypus hongkongensis; Acmaea mitra*
*Hinnites giganteus*
*Evasterias trocchelii*

1. *Nereocystis-Laminaria* (Melanophyceae) - Lacuna (= Algal) faciations (0–20 m. depth).

Algae
*Nereocystis, Laminaria, Agarum, Desmerestia, Costaria*

Animals
*Lacuna porrecta, L. divaricata, Caprella* sp.,
*Pentidotea resecata;*
*Margarites succinctus;*
*Epiactis prolifera.*

2. *Dasyopsis-Halosaccion* (Rhodophyceae) (= Algal) faciation (deeper than 1).

Algae
*Dasyopsis, Halosaccion.* etc.

Animals do not show a clear-cut special community here.

3. *Cardium-Yoldia* faciation on mud bottom.

| | | | | | |
|---|---|---|---|---|---|
| Yoldia scissurata | . . . . . . . | 10–900 | Nucula castrensis | . . . | 2–400 |
| Cardium californicum | . . . . . | 10–80 | Marcia subdiaphana | . . | 10– 80 |
| Solen sicarius | . . . . . . . . . | 10–30 | Polinices pallida | . . . . | 300–600 |
| | | | Crago alaskensis | | |
| | | | Venericardia ventricosa | . | 1–100 |

4. *Pisaster ochraceus* faciation.
An ecotone (transition) species prevalent in a narrow zone between S.P. association and the tidal community.
*Pisaster ochraceus; Stichopus californicus; Cucumaria miniata.*

II. *Strongylocentrotus-Pteraster Tesselatus* Association.

SHELFORD's general conclusions will be discussed later on. Animals present the dominant form of organic life on the bottom of the sea and of lakes only. On the soil surface they are replaced by the plants and the animals are secondary descriptive units only, the subdominants.

## 3. Description of Animal Communities: Some terrestrial associations

Very conclusive are the results of a study by Mrs. LARSEN[15] on the beetles tunnelling in the sandy shores near Copenhagen. Two sections from the seashore towards the interior show that, at least in the Skomagerslette, clean-cut associations of these beetles exist in the different biotopes of this area.

One quantitative example, containing these beetles from 1/100 sq.m. from the sandy hills of the Skomagerslette, may be added:

| Species of beetles | No. of individuals |
| --- | --- |
| Bledius rustellus . . . . . . | 73.9 |
| B. opacus . . . . . . . . . | 0.4 |
| B. arenarius . . . . . . | 0.6 |
| Oxypoda exigua . . . . . . . | 3.8 |
| Dyschirius (thoracicus + globosus) | 5.6 |
| Heterocerus hispidulus . . . . | 1.0 |
| Melanimon tibiale . . . . . | 2.9 |
| Aegialia arenaria . . . . . . | 0.6 |
| Aphodius plagiatus . . . . . | 0.2 |
| Total no. of individuals . . . | 89.0 |

Fig. 31 illustrates conclusively that most of the tunnelling beetles are able to choose actively their preferendum in a gradient of soil salinity or soil humidity. And this preferendum corresponds fairly well in all cases with the conditions of salinity and humidity prevailing in the normal habitat of the species. It would probably be premature to conclude that these are the optima for the species concerned. It may well be that these are the optima and preferenda for the local populations of the beetles and the tolerable total range or reaction basis of the species is larger. It shows in any case, however, that the individual beetles are guided in the choice of an area by environmental conditions agreeable to them. The distribution of every species in the different biotopes shows a fluctuation of its own, independent of that of the other members of the same "association". Size of sand grains had no recognizable influence on the distribution of these beetles. The predators, though also dependent upon the abiotic factors of the environment, showed a high dependency upon the distribution and density of their prey (*Dyschirius* upon *Bledius* and other Staphylinid beetles). However, as a rule, every species of *Dyschirius* shows a preference for the area of one of its hosts (*Bledius* or *Heterocerus* species; adults and larvae). On the other hand, *Dyschirius* species, generally connected with one of those hosts, may occur where no beetles of the host genera are present at all. Mrs. LARSEN found them feeding on soil nematodes in such localities. None of the predators are definitely restricted to one host species. The fact that they show preferences for certain hosts

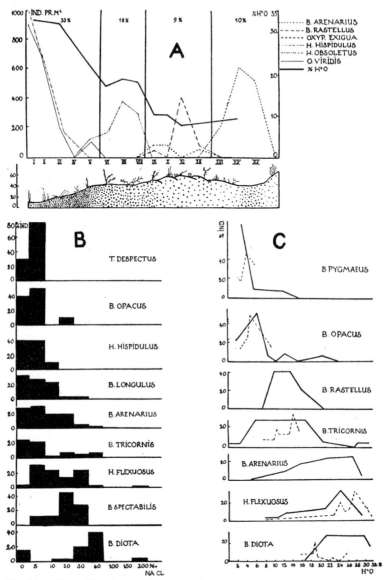

Fig. 31. Tunnelling beetles living together in the sand dunes near Copenhagen. A. Cross section through the Skomagerslette. The curves indicate the distribution of various beetles and of soil humidity throughout the different areas. The scale of the cross-section indicates the depth of the soil-water surface. B. Preference of the various beetles in a gradient of NaCl-content in the soil. C. Preference of the various beetles in a gradient of soil humidity. The broken line indicates the distribution of the beetles in their natural habitat with regard to soil humidity (After LARSEN, 1936).

**Table XXVIa.**
Life communities of tunnelling beetles in sand (After LARSEN)
*A. Section through Sibiriens Havvending (near Copenhagen)*

| Niche | % H$_2$O | % NaCl | Characteristic species | Zoocoenosis Other members | Predators |
|---|---|---|---|---|---|
| Inundation-plain | c. 20 | c. 20 | Bledius arenarius | Atheta exilis | Dyschirius thoracicus |
| Inflation-plain | Traces | Traces | No tunnelling beetles | .. | D. impunctipennis |
| | | | | | Dyschirius thoracicus |
| Sand-plain beyond the dike | 8–10 | 5–10 | Bledius arenarius | .. | D. impunctipennis |
| | | | | | Dyschirius thoracicus |
| | | | | | D. impunctipennis |
| | | | | | D. politus, D. obscurus |
| | | | | | Bembidion pallidipenne |
| Sand-plain within the dike | 12–14 | 20–40 | Bledius arenarius | Heterocerus flexuosus | Dyschirius thoracicus |
| | | | | | D. obscurus |
| Algues-plain | 18–24 | 20–60 | Ochthebius marinus | Bledius arenarius | Bembidium varium |
| | | | Heterocerus flexuosus | Hypogastrura viatica | B. pallidipenne |
| | | | | | Dyschirius thoracicus |

*B. Section through the Skomagerslette*

| Niche | % H$_2$O | % NaCl | Characteristic species | Zoocoenosis Other members | Predators |
|---|---|---|---|---|---|
| Sand-plain | 10 | Traces | Bledius arenarius | Melanimon tibiale | Dyschirius thoracicus |
| Sand-hills | 9 | Traces | Bledius rastallus | Bledius tricornis | Dyschirius thoracicus |
| | | | Oxypoda exigua | Aphodius plagiatus | D. globosus |
| | | | | Hister aeneus | |
| | | | | Bledius opacus | |
| | | | | Aegialia arenaria | |
| Plain | 18 | Traces | Heterocerus hispidulus | Oxypoda exigua | Dyschirius thoracicus |
| | | | | Heterocerus flexuosus | |
| | | | | Aphodius plagiatus | |
| | | | | Liodes ciliaris | |
| | | | | Cercyon tristis | |
| Lowland covered with algae | 33 | Traces | Heterocerus obsoletus | Helophorus nanus | Bembidion pallidipenne |
| | | | Ochthebius viridis | | |

proves that they, too, probably choose their environment following physical agreeability. It is regrettable that Mrs. LARSEN did not include in her extensive work experiments on the preferenda of the predators for soil humidity and soil salinity.

We have been studying areas of terrestrial communities in Israel for a number of years. Partial results for some areas and groups have been published[16-19]. But, on the whole, this survey is still incomplete and will remain so for many years to come. The areas (about 100 sq.m. on the average) have been chosen in typically developed plant-associations and have been under observation twice per month from sunrise to sunset for a whole year.

So far the results of this survey are rather discouraging from the point of view of the dynamic community concept and stress the independent distribution with regard to area, season, and abundance of every individual species. Fig. 32 illustrates this statement rather well.

Combining the general personal experience gained during this survey with the quantitative data obtained, we arrive at the same conclusion as STEPHEN – that zonation is a more natural way of description than minor communities. This is decidedly true, especially for the poorer life-habitats: those of the desert and predesert, the batha (= degraded maqui), tundra, antarctic, etc. In a much variegated, "broken" landscape of general high density of life, especially when rich in species, communities may easily be described, but every exact survey will show the same result which holds good for marine littoral communities of minor rank: the number of communities grows embarrassingly with every survey, and the difference between them is a permutation amongst a small number of always reappearing dominants, whereas the number of rare characteristic species is more and more reduced to monophages on a typical plant. This experience does not support the theory of equilibrated life-associations with highly integrated superorganistic structure and function.

It may be added that the phenomena of diurnal and seasonal succession, of stratification, and of vagility make the terrestrial animal communities a rather complicated study[20].

Some open-eyed botanists have come to similar conclusions. HEIM[12] describes the tropical primeval forests in their heterogeneity:

"Ça ne sont pas les quelques cinquante essences de toutes nos forêts françaises qui se retrouvent, mais un millier d'espèces ligneuses, parfois plus, qui ne connaissent aucune loi de hiérarchie. La proximité n'a qu'une règle: la lutte pour l'existence. Les agencements insaissibles n'ont qu'une cause: le désordre. En vain des botanistes sociologues s'efforceront-ils de lui appliquer les directives théoriques et dogmatiques des associations végétales, nées, sous nos latitudes,

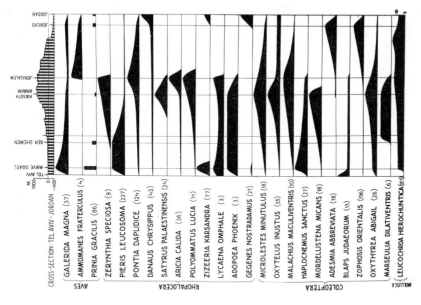
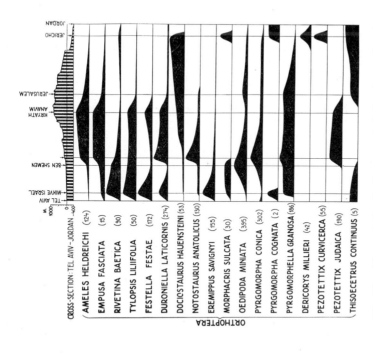

Fig. 32. The variation in the numerical abundance of various animal species in Palestine in a cross-section from the dunes near Tel Aviv to the Jordan near Jericho reveals the independent variation and distribution of every species.

des causes perturbatrices que les hommes ont imposées aux agencements physionomiques de la végétation.

"Car qu'est-ce que l'association végétale sinon une interprétation commode, simplifiée, algébrique, de rassemblements d'espèces végétales et d'individus, en vérité liés à l'action simultanée ou correlative de facteurs dont la part respective n'est pas aisée à découvrir? Pour les sociologues de l'école nouvelle, dans de tels groupements les plantes ne sont que des pions hiérarchisés, exactement comme le sont les divers pièces d'un jeu d'échecs. Ainsi est né le concept artificiel et systématique de l'association et des entités collectives dont les dimensions élargissent ou réduisent les limites de celle-ci. Aboutissement logique d'une conception fausse. Elle peut la rigueur se défendre jusqu'en deçà d'un certain degré de complication. Mais, en fait, de toute manière, l'association végétale n'offre ni la valeur ni la solidité du postulat spécifique. Nulle raison ne saurait rapprocher les unités taxonomiques des prétendues unités sociologiques. Vouloir assurer d'autre part que les relations entre les composants d'une telle association offrent la signification obligatoire qui peut s'appliquer à des partenaires liés entre eux par une action directe étroitement physiologique, à une sorte d'attractive sympathie, est conformé à un symbolisme qui trouve peut-être son explication dans l'origine germanique de telles doctrines, mais qu'un esprit qui ne puise que dans les faits d'observation ses sources de rationelle déduction ne saurait admettre sans de sérieuses reserves. Une "association" est un groupement floristique dont la relative constance statistique s'explique par celle des facteurs dont l'action conjuguée crée la biocénose, un attroupement que des influences extérieures, causes de particularités stationnelles et uniformes, maintiennent dans sa plus ou moins nette cohésion en raison de leur intervention à la fois perturbatrice, renouvellée et simplificatrice. Par le jeu de la concurrence, et sous l'influence de l'homme ou de l'animal, une constance approximative s'établie, qui ne saurait d'ailleurs éliminer l'infiltration non seulement des accessoires, mais des accidentelles, des apatrides, des imprévues... En fait, un botaniste ne devrait admettre d'association, au sens biologique, au sens physiologique, au sens littéral du mot, que là où existe une relation intime de dépendance entre deux ou un petit nombre d'êtres vivants: les notions fondamentales de commensalisme, d'épiphytisme, de parasitisme, de toutes les modalités réunies sous le terme de symbiose... sont les seules qui puissent effectivement concerner des associations... Laissons à la Nature sa complexité et à nos concepts leurs objectifs et leurs limites..."

GUINOCHET[19a] is also much more cautious in his analysis of plant associations than most plant sociologists*.

---

* See note f page 268.

BOYCOTT[20] states that British land Mollusca do not form specific associations with one another or with other animals or plants. Competition is of no importance as a limiting factor. If we arrange the Mollusca of various habitats according to the number of Mollusca living in them, we find that the kinds which live in the poorest places live also in the richest ones. The progression consists in the addition of fresh species without the elimination of those already on the list.

"*Arion ater* is probably the most ubiquitous species and often lives by itself on acid inhospitable moors. *Agriolimax agrestis* is almost as universal and *Arion minimus* not much more particular. Among Testacea *Hyalinia alliasia, H. fulva, H. radiatula, H. cristallina, Punctum pygmaeum, Helix rotundata, Colchicopa lubrica*, or *Vertigo edentula* are most likely to be found if any snails at all are discoverable. Which of these is found depends apparently on the nature of the badness and decidedly on geography. All these species are also common in the "best" places, which may show nearly 40 species in a small uniform area (e.g. old calcareous beach wood). The same is true for the Mollusca of specialised habitats such as marshes or chalk downs: the species which in some places occurs by itself or with few associates will also be found where the full fauna is present. I do not think that (with a few exceptions) a snail is ever excluded from a locus suitable for it because other snails are there already or that one species can expel another."

Their occurrence within their geographical range is therefore determined by history and the nature of the habitat, where moisture and lime are of paramount importance. The moisture is determined by environmental structure as shelter ("nooks and crevices into which Mollusca can retire to escape drought and cold and lay their eggs"). Exposure to wind is objectionable, whereas a certain air circulation is necessary. Dampness determines the amount of time a mollusc can spend in feeding and breeding per annum.

In Britain groups of snails inhabiting wet places, dry ones, and human settlements may be distinguished. The rest, about half of the species, lives in ill-defined "woodland" habitats the suitability of which varies in proportion to the shelter and lime they provide. With regard to lime there are: 1 calcifugous species, 20 calcicoles, 18 which prefer lime, and some 45 which are indifferent (within a certain range!).

The same author[21] comes to almost identical conclusions with regard to the fresh-water Mollusca of Britain. These are not dependent upon other animals except that *Anodonta* and *Unio* have an obligatory parasitic phase on fish. Nor have they any specific food. There are indications that the habitats of a few species are partly determined by competition with other Mollusca. Species as *Limnaea glabra, Planorbis spirorbis, Pisidium personatum, Physa hypnorum* are found in bad habitats only, where no other species thrive, whereas they do not occur in favourable loci.

"But in general competition is unimportant and most of the species which occur in the worst places are found also in the best. Hence each species can live in any place where its characteristic needs are met by the physical and chemical conditions. (Cleanliness of water, absence of disturbance and presence of lime are favourable). With a few exceptions, the needs of the species are so similar that habitats can be classed as good or bad for Mollusca as a whole: the richest places are calcareous rivers, lakes and canals, the poorest rapid streams, mean ponds and mountain lakes."

## 4. The Habitat Concept of Animal Communities

Repeated attempts have been, and will continue to be made, to describe animal communities by the extrinsic factors of the habitat. Type of soil, type of vegetation, average temperature or humidity, salinity, etc., range of fluctuation of the latter, etc., should always be registered. But following our present experience, those descriptions are never sufficiently neat to explain the absence of other communities and of their dominant members. PETERSEN[8] states:

"At first it was thought possible to determine those zones by the depth alone or by a characterization of the vegetation and bottom conditions, but though much has been gained along these lines we do not hereby come to the kernel of the matter, namely, the occurrence of the generally distributed animals, which alone can tell us where a certain animal community belongs, even though the depth and outer conditions may vary."

The description of animal communities is therefore an important task.

SHELFORD[13] holds a more optimistic view, also based on benthic communities of the littoral. The two subtidal biomes recognized differ in speed of water circulation, salinity, penetration of light, plankton density, and quantity of detritus. SHELFORD generally maintains that "the general hydrographic conditions (submarine climate) are more important than kind of bottom materials in determining the character of benthic communities". In his opinion the communities should form part of any physical survey.

It is an uncontestable fact that terrestrial biogeographical regions and zones may be characterized very neatly by climograms based on monthly averages of temperature and precipitation of the average macroclimate of these zones. It is the more surprising to find that the differences between the climograms are, at least in Israel, about as large in the average territory occupied by every region in Israel as are the average differences in the macroclimates of the Mediterranean (= maqui), Saharo-Sindian (= desert), and Irano-Turanian (= steppe) regions in general[18]. The importance of the macroclimatic climograms is really surprising in view of the fact that this macroclimate is purely an idealization of a large and varying range of different eco- and micro-climates and of most varying weather conditions. Certain types of soil and vegetation are often correlated with each zone.

The selective power of an unsuitable environment has been beautifully demonstrated by LUNDBECK[22] for *Chironomus bathophilus* in the Lake of Ploen. The eggs of this gnat are indiscriminately distributed over the whole surface of the lake. Projecting this surface area to the different zones of depth within the lake, out of 68 milliards of eggs, 16 milliards should drop to the bottom at 0–16 m depth, 50 milliards at 16–36 m and 2 milliards at 36–44 m depth What actually occurs is as follows:

| Depth | 6.VI | 11.VII | 31.VII | 5.IX |
|---|---|---|---|---|
| 0–7 m. | .. | .. | 52 | 489 |
| 8–16 m. | .. | .. | 2,133 | 1,045 |
| 17–26 m. | 199 | 968 | 5,967 | .. |
| 27–36 m. | 267 | 256 | 649 | .. |
| 27–44 m. | .. | .. | .. | .. |

This shows that all eggs sinking to the bottom of the lake beyond the zone of 16–36 m depth die. The intrusion into higher strata is a secondary larval migration. Animal zonation is here decidedly a selective process after indiscriminate dispersal.

The story of the plaice *(Platessa platessa)* in the North and Baltic Seas is another striking example of this type[23-27]. The plaice population of the Baltic Sea is unable to maintain itself because unfavourable hydrographical conditions (salinity, temperature) reduce even its enormous egg-production to far below the level of the parent generation.

However, sufficient as the external factors may be to give a good idea as to the general character of the environment of every zone, they are certainly insufficient to "explain" the distribution of the species. Many analyses of individual species have aimed to reach this purpose by describing the physical ecology of each species in as much detail as possible. It has not yet been possible and will probably never be so to reach this aim. Zones or conditions of optimum, favourable, unfavourable, and impossible environments have been circumscribed and hold good in rough outlines when compared with the facts. But optimal zones remain where the species is absent or has a low population level due to obscure (does this always mean: not yet analysed?) causes; there are impossible zones where a moderate population is constantly maintained, and there are zones of equal bonitation where the population level of the same species is very different. None of the combinations of factors studied in connexion with physiological tolerance have ever explained satisfactorily the local distribution and limitation of any animal species, be it in a terrestrial, soil, freshwater, or marine space. And this elusiveness in the physical ecology of an animal has its well-founded reasons.

We have left the period in which animal ecology regarded the animal as an object solely. We are learning that the responses of an animal to its environment are different from those of any non-living object. The animal reacts actively by changes in behaviour or by regulations of its physiological or morphological function or structure. A few of those physiological reactions or reconstructions are quoted herewith to explain our case. The thermo-hygrogram of an insect, i.e. the indicator of its vitality, often changes from generation to generation according to the sequence of environmental influences acting on the preceding generation, as well as on the previous development of the generation concerned (i.e. its history). SHELFORD[28] has demonstrated this for the Chinch Bug, ZWOELFER[29] for the Nun Moth, and BODENHEIMER[30] for the Red Scale. Whereas the position of the vital optimum is not shifted at all, the distance between the isothanates (= lines of equal mortality), expressing the sensitivity and tolerance of the animal to temperature and humidity, varies. The physiological resistance and vitality of the same species towards the same combination of environmental factors is therefore different in various generations and years.

Information has only recently been gathered on the shift of other physiological activities within the area of distribution of the same species. It has become a well-known fact that the thermal sum needed for the completion of one generation is often decidedly smaller on the northern border of the area of the same species, than within the centre or the southern part. Owing to this adaptation alone the animals mature in time, and in those cases where the animal is unfit to pass two winters in an immature stage negative selection is perhaps the reason for this phenomenon.

Physiological adaptation or toleration may occur in widely distributed species. MAYER[31] found in *Aurelia aurita*:

|  | Average surface temp. in summer | Limits of activity (pulsation) |
|---|---|---|
| Nova Scotia | 14° C | −1 to 29° C |
| Florida | 29° C | 12 to 36° C |

M. Fox[32, 33] reports for 7 pairs of species of marine invertebrates inhabiting southern English waters and more northern seas respectively that the oxygen consumption of the warmer-water species is greater than that of the colder-water species in their normal environmental temperatures, although the locomotory activity of the former is apparently not greater. He suggests that the similar locomotory activities of both species require approximately equal amount of oxygen, but that the non-locomotory oxygen consumption of the warmwater species is higher than that of the cold-water species. The same is correct for the cleavage of eggs of sea-urchins from England and Southern France[26]. This theory agrees with similar studies made on marine algae.

The increase of respiration-intensity of salt-water organisms entering brackish or fresh water is well known. It is expressed by the distribution of certain Crustacea which can live in fresh water only in the coldest parts of lakes, where their oxygen need is still relatively low, whereas at higher temperatures it exceeds the respiration facilities. The behaviour of the amoeba *Mayorella palestinensis*[34], which grows in zones of different depths (i.e. at different oxygen tension), in cultures of various H-ion concentrations (8.2–5.7) or of different NaCl concentrations (10.0–0.0%), is probably connected with changed behaviour towards oxygen tension. Towards acidity and towards lower salt concentrations the zone of growth falls, i.e. the amoeba grows positive to a lower oxygen tension. In concentrations of dextrose from 0 to 2% no influence on the reaction to oxygen tension is recognizable.

All these facts tend to show that in a changed environment within the area of distribution a real physiological reconstruction of the organism occurs quite regularly with regard to the local range of certain factors. The recent studies on "Abbau" diseases of potatoes confirm this viewpoint. The conclusion to be drawn is that the species is really composed of very different physiological populations or lines which are grouped geographically and conditioned by a different range of certain environmental factors, which is one of the main reasons why any absolute and really conclusive physical ecology of any animal species is impossible. For this reason, a detailed habitat description of the environmental conditions necessary for any animal community is not feasible. Whereas animal communities exist and are correlated with certain macro-climatic soil and vegetation types, no definite physical limitation can be given or be expected either for the community or for the individual species which are its members. The non-spatial ecoworld is by now replacing the concept of the area.

## 5. Empirical Species Combination, as Characteristic of the Animal Community

The conclusion to be drawn from all the above remarks is that, as a rule, each animal species has its highest density and its limits in accord with the optimal, or, alternatively, the intolerable features of the environment. Whereas the reaction basis of almost every species is somehow different from that of all other species, the areas of no two species are probably exactly identical. However, this difference may be smaller or larger and the areas of the species will vary in accordance with these differences. If a series of animal species is constantly met with in certain niches or biotopes or plant associations, this simply implies that they find here suitable or optimal conditions. They therefore thrive within this area each in his own right and not as a member of an animal association.

This does by no means exclude the fact that certain animals depend on the presence of other animals, as is obvious in the case of specific or oligophagous parasites, predators, and true symbionts. It is also obvious in the case of coprophagous scarabaeid beetles, which depend on the presence of large phytophagous mammals for food and reproduction and, again, in the case of animals developing in carrion, to the presence of which they are confined. Other species depend entirely on the presence of other animals whose deserted holes, etc., they use as a regular habitat or as breeding niches. But in all these cases the inter-related or coacting animal belongs to the environmental conditions required for existence. And in any one of these latter cases the relation is quite one-sided: the mammal may live in the absence of any scarabaeid beetle, etc. The mixed herds of antelopes, gazelles, zebras, giraffes, ostriches, etc. may likewise not be called associations, as the presence of one species is not essential for the existence of the others. However, living in the same habitat which suits their environmental requirements, they may or may not mix, according to their social temperament.

The vegetation is likewise important: in the case of mono- or oligophagous phytophages it constitutes the food or the host for development. However, it is more important as a part of the physical environment, creating microclimatic and other conditions, or breeding niches, favourable or unfavourable to certain animals.

This conception does not exclude the presence of competition. DARWIN was the first to realize that competition is strongest amongst the members of the same species. This is logical, as the niche of these individuals is more or less identical, in any case more identical and overlapping than that of the individuals of any other species. Whereas intraspecific competition is strongest, it decreases interspecifically with the increase in the difference of the reaction basis. Animals of the same group are therefore the most important competitors and often the only real competitors (ants against ants in Hawaii, earthworms against earthworms in South America, white man against red man in North America, cattle and rodents in grazing). Nor is this conception incompatible with the theory that the presence of one animal species may prevent the intrusion of another one.

It therefore seems that the animal association, its creation as well as its structure, is a chance one which is created by history and selection. The selection is negative, as it excludes species which find the habitat unsuitable and is also active, for animals are positively attracted by their senses to their preferred habitats in many cases. This becomes especially conspicuous in dysharmonic (extreme) habitats, where the inhabitants show a surprisingly high agreement in their taxisms, which is not the case in harmonic habitats.

There is no doubt that there exist animal communities of different orders, characterized by the abundance of some species, the dominants, and by the presence of mostly only a few individuals of other species more or less restricted to one community only, the characteristic species. But each member is a more or less independent member of the community, existing in it by right of its own ecoworld. The composition of the community is constant with regard to its dominant and characteristic species because of the common physiological and ecological behaviour of these species. The relative as well as the absolute abundance of each member species of the community fluctuates from season to season, from year to year, and in longer periods, parallel to fluctuation of the ecoworld.

The dominant and characteristic species of any community belong to the community, for they display similarly favourable response to the main physiographic factors present in the area covered by the community. The latter is therefore a purely statistical or empirical conception, which is useful in ecology as well as in zoogeography. As the community depends on the same factors which act on area, geography, topography, etc., it also plays an important role in any description of that environment (local environment as well as type of environment). SHELFORD[13] proposes therefore, for example, to include the community survey in any hydrographical marine survey.

Whereas the classification of the different ranks of communities is largely left to the personal discretion of every student, in principle the nomenclature and classification proposed and applied successfully by SHELFORD is to be recommended as a sound basis for further work[35], under the condition that it is strictly limited to description.

Should any type of dynamic association be recognized, the conception of the bio-community is certainly preferable to the independent treatment of animal and plant community, where vegetation or animals only appear as parts of environments. Even if associations are recognized only in the statistical and descriptive sense, descriptions of biocommunities are preferable.

For all the above-stated reasons it seems inadvisable to retain the terms autecology and synecology in their original meaning. But with some slight change they fit well into an ecology without dynamic animal associations. Autecology remains the study of the reaction basis of any animal species and attempts to get at the fitting in of its reaction basis into the natural environment, or the sequence of natural environments in which the animal lives. Synecology does cease to be the study of the integration in associations, but becomes the grouping of prevalent animals in different biotopes on the one hand, and of all possible association of any individual species with other animals and plants in nature on the other hand.

## 6. Do Coactions within the Animal Community Justify the Superorganistic Conception?

It would be premature to conclude that within animal communities, zoocoenoses or biocoenoses, every species reacts as if it were alone and isolated. We know definitely that individuals of the same species, as soon as they surpass a certain intraspecific density, decidedly disturb one another, which is expressed – if not alleviated by mortality or emigration – in lower fertility, under-nourishment, shortened longevity, etc. The sum of individuals of any one species within one area may show the self-regulation upon the setting in of intraspecific population pressure. The mutual pressure (= interpressure) of any two species inhabiting the same community is greater when the two are more similar in their ecological demands and niches, and smaller when their ecological needs overlap slightly or not at all. In these cases it is not a question of mutual help but of mutual toleration, no balance existing between those species. The numerical ratios depend entirely on the environmental conditions and the reaction basis of these species[37].

The coactions of animals are mainly on the food basis. The weight of all predators must always be much lower than that of all food animals, and the total weight of the latter much lower than the plant production[36]. Weight-ratios of 1 : 10–30 have been found amongst those groups, i.e. 1 predator: 20 food animals: 400 vegetable matter. But although reproduction goes on at a more rapid rate in the prey than in the predator, as a rule, owing to the smaller size of the former, it is devoid of any meaning, and a real dynamic evaluation of numeric relations of the different layers of the food pyramid of any animal community has not yet been obtained. It does not seem just to conclude from the mere maintenance of the food pyramid that intrinsic regulations occur within the community.

For any individual species we have as many examples of coincidence of fluctuation of prey and predator as of those which show a complete independence (see Fig. 33). Whereas a predator is able to maintain itself permanently only in those communities where the fluctuation of the total normal food never falls below the level necessary for its maintenance – even during the poor season – it is obvious that the importance of absolute food quantity present is of primary regulatory importance. Wherever the predator preys too much or overgrases, he destroys the life community on which his existence is based. The permanent maintenance of large territories per family is therefore a common feature among small and large

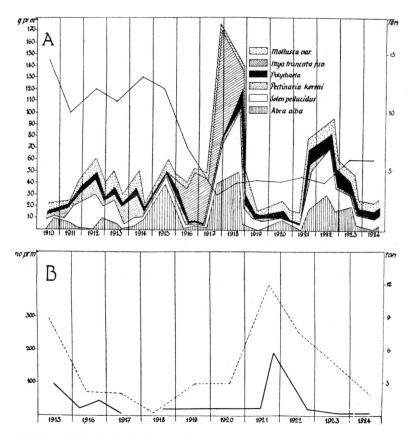

Fig. 33. Fish food and fishery yield in Danish seas. A. Quantity of fish food (gm. per sq.m.) and yield of plaice fishery (in tons: single line) in the Thistal broad during 1910 to 1924. B. Number of *Solen pellucidus* (per sq.m.) and yield of Flatfish fisheries in the Fur Sound-Risgaard broad during 1915 to 1924. (The curves A for 1915 and 1922 deviate slightly from the original of BLEGVAD)[38].

predators. After this behaviouristic exclusion of food competition its regulatory function is excluded to a certain degree. The question may be raised – it cannot yet be answered by the scanty material at our disposal – as to whether the absolute limitation of food is a common and important limitation in normal environments. Almost all facts available point to the negative. It becomes more and more obvious that limitations are primary within the animal itself. A most illustrating lesson in this respect is the sequence of Planarians in springs in Central Europe. Experiments show[39]:

| Species | Temperature in nature | Optimal temperature | Instantaneous death | Habitat |
|---|---|---|---|---|
| *Planaria alpina* | 0–10° C | 3–7° C | 30° C | The origin and coldest part of the spring. |
| *P. gonocephala* | 12–20° C | 15° C | 32° C | The later and warmer part of the spring. |

Temperature has a determining effect as far as it determines the general tonus of activity. *Planaria alpina* does not differ by much in its absolute temperature tolerance from *P. gonocephala*, but above 10° C it grows lazy, whereas its food exigencies are higher. It starves itself as consequence of this discrepancy in physiological behaviour and nutritional needs. The reverse is true for *P. gonocephala*, the activity of which is low at lower temperatures and below the level of metabolic maintenance as well as below that of active feeding. The mutual replacement of these Planarians is therefore not based on competition and no intrusion into the area occupied by each *Planaria* occurs, even if only one species would occur in one spring.

Coactions (interactions between organisms) occur in large numbers within any animal community. Organisms create conditions for the development of other species, even apart from the food problem. But is does not seem probable today, that these coactions make the statistical or geographical community a real one, a system of labile equilibrium of species interacting in the most manifold ways and integrating the members of a community into a real superorganism.

The supra-organismic concept has found its most enthusiastic support in two fields of ecology, those of limnology and social animals.

S. A. FORBES[40] wrote as long ago as 1887: "(The lake) forms a little world within itself – a microcosmos within which all the elemental forces are at work and the play of life goes on in full, but on so small a scale as to bring it easily within the mental grasp." And A. THIENEMANN has unceasingly advanced this idea: "Every life community forms together with the environment which it fills, a unity, and often a unity so closed in itself, that it must be called an organism of higher order," even if all life communities form part of the great vital space of the earth as a whole.

THIENEMANN[41] has often stressed the supra-organismic character of the lake from a dynamic, or, better, from a physiological point of view. Considering the total circulation of organic and inorganic matter within a closed system such as a pond or a lake, a certain analogy to the metabolism of an organism may be discovered. But this circulation of matter in a closed, autarctic system – apart from aëration and sunlight, common to almost all life communities – is nothing more than an analogy. It simply means that the nutritional salts are used for the production of the autotrophic flora, which is being used in turn by smaller animals which serve as food to fish. All remnants are decomposed by bacteria, and their matter again enters the circle of production either as detritus, the food of detritus-feeders or as nutritional salts utilized by the autotrophic flora. This system is therefore not to be compared with the metabolism of an organism which acquires purposely and selectively the materials necessary for the maintenance of the organism from outside, and which has a complicated physiological system of internal dissimilation, assimilation, transport, and distribution of matter. One phase of the circulation interrupted, the organism breaks down. However, in the lake, life, perhaps in the form of another life community, would be maintained if no fish were present or if bacterial decomposition were quite incomplete.

The only conclusion to be drawn is that here also the relations of the food pyramid are maintained. An empty food space within a harmonic environment always calls for a filling up with suitable organisms from the neighbouring biotopes. The mere fact of the maintenance of the food pyramid cannot be regarded in this case as a proof for supra-organismic organisation of the bio-community.

A strong additional argument against this concept is the mode of establishment and longevity of such a biotope which depends entirely on the stability of the environment. In the stable environments of the abyssal, pelagial, benthal zones of the sea conditions are fairly stable over long, sometimes even over geological periods, whereas the littoral community exposed to the rise or fall of the sea, the erosive effect of the tides, etc. is entirely unstable. Once the conditions change, the life communities follow quickly.

SHELFORD[13] states for the *Balanus-Littorina*-biome on rocks as well as for the *Macoma-Paphia*-biome on sandy beaches, that they develop within a few months. "All the communities studied are characterized by short life-histories for the representative species

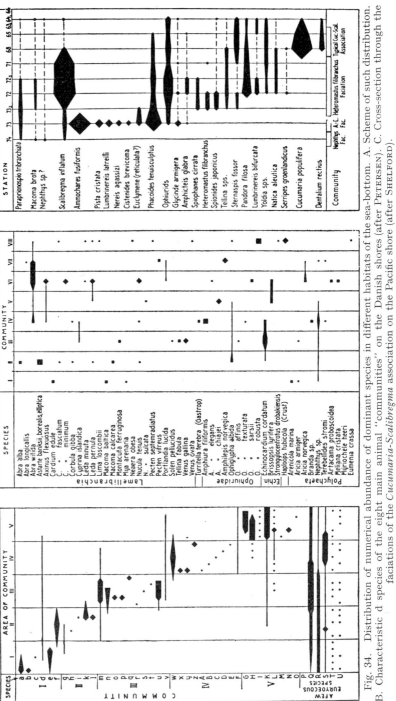

Fig. 34. Distribution of numerical abundance of dominant species in different habitats of the sea-bottom. A. Scheme of such distribution. B. Characteristic d species of the eight main animal "communities" on the Danish shores (after PETERSEN). C. Cross-section through the faciations of the *Cucumaria-Scalibregma* association on the Pacific shore (after SHELFORD).

and rapid replacement; hence, also rapid development and quick response of community composition to minor changes in external conditions." The identification of the bio-community of the lake by metabolism, mode of establishment, and longevity (dependent also on changes in the environment produced by the activity of the organisms) is thus mere analogy, but not homology.

In social animals the beginnings of integrations may lead to a higher degree of real supraorganismic organisation. We will omit here any discussion of man. In many monospecific communities, such as the families of social insects, no doubt it is possible to attain a high level of supraorganismic integrations[42, 44] with regard to division of labour and its adjustments under experimental conditions, mode of reproduction of the families, development and longevity, social temperature regulation, and so on. These integrations are in many cases so far-reaching that they are, so far, the only category of life communities for which a real homology to the organisation within multicellular organisms can be established.

These statements are entirely true for terrestrial communities too. After draining a moor, we observe the speedy replacement of the bog-biocoenosis by a transition of series of plants and animals up to the establishment of the new community. This transition lasts for 5–20 years, depending on the degree of drainage and final character of the area. But as soon as the soil has grown "mature" and becomes stable, the new community is established within 1–3 years. This means that community changes depend on the rapidity of the environmental changes solely. If a new environment is established rapidly, the new animal community is also rapidly formed. If the transition is slow, the formation of the community is also gradual and slow. The community does not undergo a natural growth-cycle in which successions of different stages of maturation supersede each other ending in death, but its growth and death are entirely unorganismic, depending solely on the permanence and stability of the physical environment. The same is true of their longevity, which is not dependent on intrinsic (physiological) factors but on extrinsic (environmental) ones purely; i.e. longevity in the organistic sense does not occur.

The fact that plants and animals of any community may change the environment and thus create new conditions for another community of the same climax-series, i.e. the phenomenon of succession of series within one climax-association, is entirely independent and does not interfere with our argument, but, on the contrary, confirms it. Series of successions are conditioned by change of environmental conditions, and each member of a series enters, flourishes, and disappears from this succession on its own accord. It is the human mechanism of thought which chooses the dominants selected at certain intervals in the succession sequence as indicators of community shift.

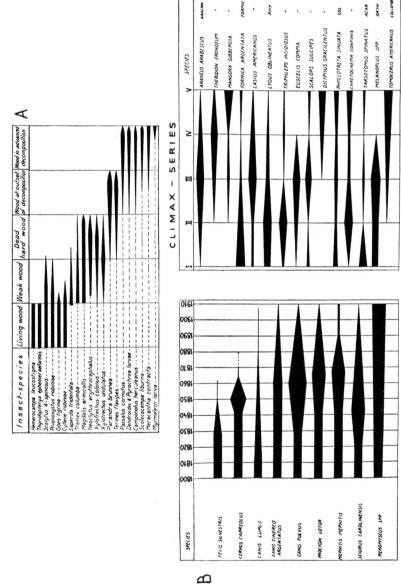

Fig. 35. Abundance changes in terrestric animals. A. Self-destructive insects in dying wood in Illinois (after SHELFORD). B. Shift in abundance of nine mammals of Illinois following human settlement (after WOOD). C. Insects of the climax-series of the Red Oak – plane tree wood (greatly schematized, after SMITH).

The animal community as an empirical or statistical conception is compatible with the descriptions, deductions, and conclusions of all those workers who have made surveys of their own in the field. This conception is compatible with the idea that the vegetation and the fauna of any landscape have general traits in common. In a desert where lack of water in any form is the dominant factor all life is dominated by this one factor. The same is true of animals living in the abyssal, pelagial, littoral of the sea or in the tundra or a tropical rain-forest. The "genius loci" (Caradja) of each major habitat is expressed by common traits in sociability, breeding-stratum, colouring, body-size, etc.

This conception, however, is incompatible with that of the animal community as a mystic holocoen (FRIEDRICHS)[44] or as a dynamic superorganism as long as it is not founded on facts but on intuition only. The author is inclined to believe on the basis of the general development of ecology that the final solution of the problem of the real character of animal communities will be a compromise. The numerous processes of coaction within such a community are mainly quite primitive and these only occur where wide areas with relative few species of animals and plants occur.

We may define animal communities tentatively as a combination of (plants and) animals, recurring in an approximately similar composition, with regard to the dominant and characteristic species at least, wherever similar ecological conditions exist within the same zoogeographical territory. Identical habitats in different zoogeographical regions show a surprising analogy and homology in the composition of their life-communities. This similarity is the outcome of selection: negative selection by the environment, positive selection by similar taxes and behaviour. Within the animal community occur interactions between animals and plants, but even beginnings of any real integration of the members of the biotic community, the biocoenosis, to a superorganismic structure are extremely rare (and then monospecific). As a rule, each species exists within the community on its own right, which is expressed by the area of each species conditioned by its own reaction basis, different from that of all other species. No active cooperation occurs in the animal community, but at the utmost, a certain mutual tolerance of those species, the niches of which overlap to some degree.

The animal community is therefore a useful empirical and statistical conception, facilitating the description of animal life in various habitats, but not a dynamic, superorganismic reality. Considering the lack of organismic behaviour, for example, with regard to growth, metabolism, longevity, which the biotic community shows, the application of the term sociology (as used in botany) is out of place.

It is, in general, a weak point of ecology that it is mainly based upon analogies between two sets of phenomena. What is really disturbing is the ease with which new analogies are built up, once the accidental character of any such analogy is discovered. Physiological foundation by experiment or by counts of area-samples are the first beginnings for the development of proper and specific methods in ecology, both still far too rarely applied.

The same is, of course, true for the term "self-regulation". We have applied it repeatedly for populations of single species in order to refer to the density-dependent processes which prevent a steady overpopulation (population high) of any species under all conditions. This kind of self-regulation is intimately correlated to our interpretation of biological equilibrium.

It is, naturally, very different to apply the concept of self-regulation to life-communities co-inhabiting areas or "biocoenoses". In every natural habitat – more or less undisturbed by man – a maximum population (under the given circumstances) of all species available co-exists. This situation we call poetically the "harmony of nature". It is mere intuition to claim that this "harmony of nature" is more than mutual tolerance, to make it the outcome of complicated self-regulations and integrations, interpreting the life-communities as a true homology of an organism, a true super-organism. Anything of this kind has still to be proven.

We are fully entitled to use the term "harmony of nature" in the given sense, because man has created increasingly growing areas of "disharmony of nature" everywhere he directly or indirectly has brought about a change making the natural rejuvenating forces of the soil, the vegetation and the fauna inferior to the destructive forces of erosion, overgrazing, overhunting etc. This means that where man has created conditions under which natural rejuvenation cannot overcome the destructive processes by which the soil is denuded of its fertile layers, the vegetation is replaced by hardy plants of less coverage interfering with the continued use of the area as grazing grounds of domestic stock, the game may eventually be exterminated.

The harmony of living nature has occasionally roused an almost religious enthusiasm amongst ecologists, almost comparable to that of the age of JAN SWAMMERDAM when biological and microscopical research was regarded as the study of the wonders of God's creation. German ecologists have compared the supraorganismic hypothesis of the bio-communities to listenening to the "symphony of the spheres", a well-chosen simile stressing the intuitive character of this theory. It is obvious that no scientific method can arrive at a perception of the symphony of the spheres. It is in this sense, and in this sense only, "that we would gladly accept the accusation of being a non-musical person, unable to perceive the music of nature's composition.[45]".

To this S. L. Tuxen remarks: "To refute beforehand what is not "probable today", seems to me to be a sterile procedure; even if these coactions were mere ripples at the surface, it is more useful to assume that they are of farreaching importance, if we are to investigate them, than to reduce them at the outset to mere trifles. Altogether it will be more useful to many investigators to have a picture, a "vision", before them before they set to work, even though it may be wrong, than to work in such a way that each result merely gives rise to the next problem ... It seems to me that if "musical" in this connection is to be regarded as identical with the ability to enter intuitively into the subject-matter, it means a serious defect to admit that one lacks this ability."

Thus, Tuxen is one of the few modern ecologists who does not pretend that the supraorganismic biocoenosis is a proven fact, but honestly calls it a vision stimulating synecological research.

The crux of the problem is the wide-spread lack of epistemological understanding in biological thinking, already deplored by Lotka[46] and many others. No partisan of the empiric school has to our knowledge ever denied the possibility of a supraorganismic structure of bio-communities. This school has only stressed that at the present state of our knowledge the factual basis of the analysis of the "web of nature" is entirely inadequate for a general synthetic solution. Scientific method proceeds from description, observation, measuring, counting, weighing and similar methods to an induction which is evolved from the sum of available facts and observations. From such a purely methodological point of view it is impossible to give any picture and decision on the inherent structure, if such exists at all, of the bio-communities. We agree, however, with Tuxen that any heuristic theory is better than nothing as guidance for research and scientific analysis. This has been realised from Aristotle[47] to Singer[48]. The former stated: "Such appears to be the truth about the generation of bees, judging from theory and from what are believed to be the facts about them; the facts, however, have not yet been sufficiently grasped; if ever they are, then credit must be given rather to observation than to theories, and to theories only if what they affirm agrees with the observed facts."

We well remember the enormous stimulating effect which the theory of the biological equilibrium between hosts and parasites[49] had upon the development of ecology as well as of applied entomology, and cannot therefore belittle the value of any general heuristic theory. What we refute is the claim of any such theory, and especially one still so vague as the supraorganismic structure of bio-communities, to truth. We accept it as a heuristic principle, stimulating and fascinating, which, like all generalizations followed up to their last consequences, will not only elucidate its own degree of truth or fallacy, but will bring out a great number of facts and relations which otherwise would not so speedily have been discovered. We gladly accept thus the supraorganismic community concept as a valuable heuristic principle but we refute its claim to be an established fact. The general consensus of the ecological authorities of our day shows that their intuition, built upon the subconscious digestion of the observations of their lifetime, strongly suggests and supports this view. Only history will show if this consensus belongs to the true intuitions of RUSSELL or is a fashion of our time or is eventually based upon the unifying and simplifying mode of operation of the human mind.

The human mind has an inborn tendency – we might perhaps even say need – for the simplification of the complex, for the unification of the diverse, of the phenomena of nature as well as of other fields of human thought. The supraorganismic biocoenosis is merely an anticipation of possible results. It is believed, yet not proved. This does not mean that it is false or untrue, but only that so far it is an aprioristic anticipation of our mind which may be right or may be wrong. KANT points out that such anticipations of our mind are one of the important ways of cognition of the human mind. If consensus of the authorities would be a proof of truth, then it is right. We have, anyhow, no reason not to believe in the supraorganismic character of life communities, as long as it is clear that it is based upon intuition, and not upon scientific method. The only demand and restriction must be an agreement that it is based upon another road of cognition than observation and experiment.

The pure scientific method of induction cannot, as RUSSELL points out, reach the realms of higher knowledge, of complete synthesis of general principles. True scientific synthesis comes not from induction but from intuition or deduction. The latter remain, however, theories which require confirmation by inductive experience. Their intrinsic value depends upon the number of non-connected (prima facie) phenomena which they explain and, of course, that – even if part of the pertinent phenomena are left unexplained – they shall not be in contradiction with any one of them. Outstanding illustrations of valuable theories are the transformation of species, the chromosomal base of heredity, the cell-theory, none of which is today – except perhaps the first – without serious criticisms and discussions. Yet all three of them have in a limited or changed form a good chance to survive our generation. That the same can be said about the supraorganismic biocoenosis is much more doubtful.

The present-day phase of this theory does not give any functional explanation beyond mere analogies. There is no link between the observed facts and the theory. None of the assumed links surpass fractional chains of relations (food-chains, life-cycles, foodpyramid, etc.). Too simple interpretations are certainly bound to failure.

Interspecific relations within the biocommunity definitely do not form a well-balanced system in which every minor change of composition has far-reaching consequences for the structure of the entire community, such as the healing even of a minor wound signifies for the entire organism. Buffering by automatic adjustments as well as intercompensatory tendencies[50] occur as widely established principles. They counteract the effect of minor changes within the bio-community and tend to give it the observed stability. Yet similar phenomena are known from physical and chemical systems and hence do not call for explanation by supraorganismic structure.

J. Pelseneer[51] tries to demonstrate that there exists no general scientific method at all and that scientific progress is maintained by the artistic temperaments amongst scientists. The background of this well pointed thesis is, of course, the same which Russell expressed in another way: that mere induction does not lead to new guiding principles, but only intuition and deduction. This brings us to our main thesis: There are more scientific roads to knowledge than one. Scientific induction gives us well-measured and secured facts. The theories which are the soul of science are, however, gained by intuition and deduction. Both roads are admissible and of basic value. The only mortal sin is to anticipate theories as established facts when the inductive material may suffice to stimulate an inspiring vision, but do not – or not yet – form a solid base for truth or even for a high degree of scientific probability. Tuxen has said truly that it is a poor science which renounces visions, but it is a still poorer science which commits the mortal sin to accept vision as science. Science advances only when it passes between this Scylla and Charybdis into the open sea of balanced judgment between both these methods.

# VI
# THE INTERACTION OF ENVIRONMENT AND HEREDITY WITHIN THE ORGANISM

## 1. Actual Relations between Ecology and Genetics

"To ask whether heredity or environment is the more important is like asking whether the area of a rectangle is more dependent on its base or its altitude[1]." This clear expression of an American student characterizes the polar relations between the two branches of biology, genetics and ecology. Whereas no anterior, northern, etc. pole is imaginable without the presence of its counterpart, a posterior, southern, etc. pole, no organism is imaginable not subject to both major forces, those of heredity and those of environment. Both forces work on the organism, which in turn responds to both by action and reaction, and in this way its many potentialities are realized.

Considering this close essential connexion between constitutional and environmental factors within the organism and their permanent mutual interrelation, close co-operation ought to be expected between the students of both branches of biology. Since the development of population genetics some degree of cooperation has slowly evolved, even some interchange of opinion. There is still, however, a lack of full cooperation, apparently the consequence of different stages of development of both sciences, resulting in widely differing maturity and widely differing methods.

The problem of heredity has been recognized as one of the basic problems of biology since the days of LAMARCK and DARWIN. Variations in plant and animal were studied and statistical methods have been applied to them since the days of JOHANNSEN. The rediscovery of the Mendelian laws introduced quantitative analysis and prognostic methods of individual hereditary characters. However, genetics really matured only after MORGAN's ingenious foundation and quantitative demonstration of gene structure within the chromosomes of *Drosophila*, and after the foundation of development-mechanics by ROUX and SPEMANN, which approaches the another angle. Now it became possible to obtain clear-cut answers to clear-cut questions by experimenta crucis, provided the genetic structure of the parents to be crossed was known. But this type of experimentation, which always yielded quantitative results proving or disproving a theory, eventually after a suitable change of the original plan of experimentation, resulted in a self-certainty which is still characteristic of the genetical way of thinking of our time. The geneticist refuses any conclusion which cannot be demonstrated and proven by experiment. The experiment is certainly one of the basic methods of biological argument, and the results of genetical experimental methods in our generation certainly justify the pride of these workers; but they forget that negative experimental results bear a great deal less weight in biology than positive demonstrations. They forget that the development of almost any branch of biology shows that nature very rarely chooses one principle only, one solution only for any morphological, functional, or psychological problem.

Animal ecology is a relatively young science in comparison with its older sister. Conclusions drawn by analogy or coincidence still play a major part in our thinking. Shifts of population density coinciding with parallel trends of temperature, precipitation, etc., are regarded hypothetically not as analogous, but as causally connected. Should the inadequacy of any such connexion be demonstrated, a new theory by analogy would, as a rule, easily be formed. The growing empirical registration of facts, trends, and analysis under different conditions gradually improves the conclusions drawn and makes them more and more reliable. The introduction of physiological laboratory methods opens important ways of checking the theories gained by analogy. But the ecological interpretation of facts will always be less reliable than the genetical one. In genetical experimentation the genetical structure (at least with regard to qualities concerned) is quantitatively known, if not at the beginning then at the end of the experiment. Environmental factors are generally kept uniform or are at least well known. In ecological experimentation we may study the influence of one factor or one combination of factors – other conditions equal or known – experimentally.

In nature, however, the interaction of all factors changes the value of the experimental result to such a degree that a conclusive deduction is difficult. Moreover, whereas the ecologist experiments with constant factors or with known changes in a limited set of factors (temperature, light, etc.), the influence of the physical environment is often different in nature. Moreover, the various successive stages of an animal often have very different optimalzones and show a varying sensitivity towards external influences at different ages. The environmental factors – biotic and abiotic, endogenous, exogenous – often have a major influence during one or two very short stages of development, and if this stage (often a few days only out of a long-lasting egg-stage) is not included in the experiment, only inconclusive results are obtained.

And, finally, in almost every case the genetical structure of the population concerned is usually unknown. Who will reproach the geneticist for preferring his pure, clear-cut kind of thinking to this state, where methods and results are obviously still in an early Mendelian stage, and where genetic studies are needed as an urgent completion of any monographic research? However, his manifest derision of ecology has still some other grounds. He fears the invasion of the ecological point of view into his field. When this invasion becomes more and more urgent, he knows that the period of the Olympian mind is over and fears that the consequent revolution of genetics will rob him of the almost unlimited certainty of conceptions and results which are so satisfactory to him today.

In view of these unsatisfactory relations between the two sciences we shall here try to underline the importance of genetic research for ecology, and it is hoped that geneticists may soon try to review the importance of ecological research for genetics.

## 2. Heredity of Important Ecological Characters

We shall first turn our attention to a series of genetic facts closely concerning the student of ecology.

Environmental factors commonly influence morphological characters of the organism. The number of facets in the eye of *Drosophila*, the number of bristles on a certain part of this fly, the tail-length of rats, the colour of mammals and insects are decidedly influenced by temperature. Every change in environment may influence the percentage of crossing-over and the mutation rate.

However, fundamental changes in vitality and other physiological functions very often accompany morphological changes, and this is of paramount importance to the ecologist:

(1) Longevity is based on genetic constitution. In crossing two strains or races longevity is often longer than in either of the parents. For *Drosophila* PEARL[2] reports:

| P | | $F_1$ | $F_2$ Tr. x W | $F_2$ W x Tr. |
|---|---|---|---|---|
| Wild | Truncate | | | |
| 40.5 | 26.9 | 47.8 | 32.8 | 31.1 days of adult life |
| 30.5 | 18.5 | 46.6 | 25.9 | 27.3 days of adult life |

In the 1st filial generation duration of life is considerably longer than in either parent, while in the 2-nd filial generation it is intermediate (hybrid vigour).

The crossing of different races of the parasitic wasp *Spalangia*[3] yields:

| | Longevity of | Egg-production of |
|---|---|---|
| *Spalangia orientalis* . . | 15 | 75 |
| *S. sundaica* . . . . . . | 27 | 170 |
| *S. orientalis* x *S. sundaica* | 30–2 | Up to 240 |

(2) *Spalangia* also served as an example of constitutional change of fertility. In *Bruchus quadrimaculatus*[4] similar observations have been made on different mutations.

| Phenotypic colour | Eggs per pair | hatched % | No. of pairs | | Index of total fertility (100 pairs) |
|---|---|---|---|---|---|
| | | | Fertile | Sterile | |
| Wild . . . . . . | 90 | 70 | 70 | 0 | 6,300 |
| Red . . . . . . | 48 | 80 | 70 | 0 | 3,840 |
| Red maculated . | 42 | 80 | 70 | 0 | 3,360 |
| Black . . . . . | 40 | 80 | 70 | 0 | 3,200 |
| Grey . . . . . . | 13 | 35 | 36 | 34 | 514 |
| Wild x grey . . . | 18 | 69 | 11 | 10 | 524 |
| Grey x wild . . | 24 | 44 | 21 | 0 | 1,056 |

TENENBAUM[5] has demonstrated for the Israel race of *Epilachna chrysomelina* that reduction in size of a system of elytral spots goes parallel with heavy depression in vitality and fertility, being therefore an easy indicator of far-reaching physiological reconstruction of the body.

Homozygotic grey-coloured Schirazi Caracul sheep are always lethal in the first year after birth. Physiological disorders in connexion with lab-ferment secretion are connected with this mutation[6].

(3) Length of development is also influenced by constitutional changes. In *Ephestia kühniella*[7] in the mutations "red eyed" and "black scaled" the following statistically significant differences have been observed:

| Mutation | Duration of development | $M \pm 3$ m. |
|---|---|---|
| Black eyed . . . . . . | 90.8 days | 90.3 – 91.3 days |
| Red eyed . . . . . . | 92.5 days | 91.7 – 93.3 days |
| Wild coloured . . . . . | 87.6 days | 87.1 – 88.1 days |
| Black scaled . . . . . | 86.5 days | 85.8 – 87.2 days |

(4) Every change in constitutional vitality has farreaching consequences. In *Drosophila* the stable maximum population to be reached within a stable, closed environment differs with each mutation[8]. Wild *Drosophila* populations stabilized at 938 individuals in a pint bottle, those of the mutant quintuple at 319 only, and those of the mutant vestigial at 345.

The wild type is vitally stronger than almost all known mutations. This is confirmed by the fact that in mixed experimental populations of *Drosophila* nearly all mutants are eliminated by the wild form after a very few generations[9].

(5) The preference temperature of mice is also subjected to Mendelian heredity[10].

(6) Genetic constitution is also connected with disease resistance, e.g. as GOWEN[11] has demonstrated in strains of mice. The constitutional part in survival was after inoculation of pseudorabies in various strains:

S. 8.4%; Wf. 55.4%; Ba. 11.5%; Sch. 22.6%; sil. 52.2%.

A few of the recent researches on mutability demonstrate the importance of coordinated ecological research. The frequency of mutation in nature seems to fluctuate, but not seasonally. SPENCER's[12] table shows observations in different species of *Drosophila*.

It is still unknown as to which environmental factors may influence or inhibit those mutations. Cosmic radiation is probably much too weak to have any influence on them[13].

We have mentioned before that mutation-rate is influenced by temperature, but the rate at which mutability rises is about twice

| Length of period | No. of mutations |
|---|---|
| 28th month | 25 |
| 35th month | 29 |
| 38th month | .. |

as great as that of speed of development (except in highly mutable genes)[14]. Other factors as well as chemical stimulation or exposure to radiation, etc. may also increase the mutation-rate or perhaps even mutation-tendency. In radiations, wave-length has almost no influence, but the frequency of mutations is proportional to the dosage, regardless whether this dosage has been given in a high or low concentration (i.e. quickly or slowly) with or without interruptions. Whereas the presence of most genes within the cell is obligatory in all stages of development, their specific influence on certain processes or periods of development is often effective within a limited period only. This is not always the case.

In glass-winged *Ephestia*-moths part of the wing-area loses its scales[15]. Position and extent of the areas as well as percentage of individuals showing this character depend mainly on temperature. High temperature increases the extent of "glass-wingedness", but it is of no consequence whether the exposure has taken place at the end or at the beginning of development.

Sensitive periods occur for certain characters in normal development probably oftener than is realized today, perhaps because normal development has provoked special research in this respect to a smaller degree. But the response of quite a series of mutant characters to external stimuli (e.g. temperature) is certainly different in different stages of development. In *Drosophila* such results have been obtained for characters as short wing, scute-1, mottled-eye, vestigial, etc.[14, 15].

A concentration of dark rodent populations has been observed in the humid soils of the moors of N.W. Germany and in certain other limited areas of that country[16]. In *Apodemus sylvaticus* this darkening is caused by the dominant umbrous gene U which appears most commonly in the moors of Oldenburg and Friesland, and which becomes progressively rarer in areas where the soil is not humid. On the other hand, it is well known that darker fur colour is also a phenotypic reaction in a cool and wet environment. Apparently, a humid environment favours the full development of the darker chromic pigments and of their enzymes in the hair as a phenotypic manifestation. In most cases, a few breedings, cross-breedings and back-crossings will suffice to clear the genetic background in every specific case. It is regrettable that few such experiments have so far been made. The problem of where the dark colour is an ecotype and where it is an ecophene is certainly important enough to justify extended research in this direction. The next problem to be studied would be, if the darker ecophenes have selective value, either because they have a higher vitality in moist environments, or because they eventually produce protective colouration there.

The occurrence of malign tumours is very rare in wild rodents, as most of them die before the age when cancer appears. In females more than 500 days old, however, we observed mammary carcinoms in 15%. The only female which showed a malign tumour (mixed cell sarcoma) at the age of 320 days was the offspring of a female which had had a mammary carcinom at the age of 745 days – which fact hints a genetical basis for the disease[17].

The knowledge of lethal genetic combinations is often useful for the understanding of sudden population changes as a consequence of changed sex ratio or changed fertility. It is important to know that almost every mutation is a minus mutation, lowering the general vitality and often the fertility or other important processes. Selection in combination with the conservatory forces of heredity tends to maintain a population genetically stable in any given habitat, as long as its environmental factors remain more or less constant. The chance that vitally potent mutations may develop and form new species is almost only a question of the length of geological periods. This has been clearly demonstrated by KINSEY[18] for North American Cynipidae.

### 3. Population Genetics

The new field of population genetics is entirely plausible on the basis of neo-Darwinistic assumptions, on the oft-repeated observation that gene arrangements in animal populations follow certain rules. This is, for example, quite obvious from some observations of T. DOBZHANSKY[19] on some genes in the third chromosome of *Drosophila pseudo-obscura* in a line from the West to the East of Southern U.S.A.:

### Table XXVII.

Arrangements of certain genes in the third chromosome of *Drosophila pseudoobscura* in a cline in the U.S.A.

| Locality | Genes | | | |
|---|---|---|---|---|
| | Standard | Arrowhead | Chiricahua | Pikes' Peak |
| Santa Barbara, Calif. | 46 | 21 | 17 | – |
| San Jacinte Mts., Calif. | 39 | 26 | 31 | – |
| Western Death Valley | 31 | 52 | 16 | – |
| Eastern Death Valley | 17 | 71 | 12 | – |
| Prescott, Ariz. | 11 | 79 | 9 | 1 |
| Flagstaff, Ariz. | 1 | 97 | 1 | 1 |
| Mesa Verde, Col. | – | 100 | – | – |
| Raton Pass, N. Mex. | – | 80 | 1 | 18 |
| Trans Peces, Tex. | 1 | 36 | 4 | 54 |
| N.C. Texas | – | 22 | – | 70 |
| S.C. Texas | – | 12 | – | 70 |

Small, isolated local populations of the fly may show much greater differences, even from closely neighbouring sites, than is apparent from this analysis, which gives average conditions and reveals a cline. Similarly significant are the seasonal differences occurring in the same chromosome of the same species. This was in 1939, e.g. at San Jacinte:

| Date of Collection | Genes | | | |
|---|---|---|---|---|
| | Standard | Arrowhead | Chiricahua | Others |
| 24.IV | 49 | 33 | 15 | 3 |
| 13.V | 31 | 36 | 28 | 6 |
| 21.VI | 33 | 37 | 27 | 4 |
| 19.VIII | 38 | 33 | 24 | 8 |
| 19.IX | 49 | 21 | 24 | 6 |
| 21.X | 56 | 25 | 14 | 6 |

DOBZHANSKY interprets these seasonal changes as the consequences of population contractions during the winter.

A stable gene equilibrium would be reached in every population in a more or less stable environment. The environment as well as the populations, however, are in a permanent flux. The dynamics of gene selection in natural populations have been well analysed by S. WRIGHT, J. HALDANE, R. FISHER and others. The pressure of recurrent mutations, selection pressure between individuals and populations, the effects of isolation and of population size are of primary importance for gene fixation within a population.

The one-time occurrence of a single gene would by itself be of little importance for the population. In most cases it would be simply lost. It is only the regular recurrence of mutations of the same gene, of the same character, cropping out again and again, that can increase the frequency of a given gene within the population.

This point is the more important as mutations are partly reversible processes, thus inducing a certain number of the new genes to disappear and to pass into the type population during later generations. The selection pressure of any gene is not only a function of its own frequency, but of that of all the other genes. Selection and mutation oppose each other, and their joint diagonale of forces determines the equilibrium in gene frequency. The conditions under which mutation pressure is likely to dominate appear to be rather restricted. Even a very slight selective advantage – and such seems to occur often – would be more important, as a rule. Nonetheless, in very small populations selection pressure becomes ineffective, while mutant pressure is not effected.

Population size is decisive for the chances of mutation spread, apart of those of lethal genes, as it determines the chance of an independent outcrop of the mutant, and subsequently the frequency of its re-combinations. The store of mutations is, of course, larger in big than in small populations. In small populations, however, recessive and rare mutations have a higher chance of survival. A good illustration is *Drosophila hydei*. This tropical species has become widely established in the cities of the U.S.A. Its rigidly isolated city populations undergo heavy contractions every winter, followed by a rapid increase in summer. As a result of these great seasonal fluctuations no genetic equilibrium is ever established. Outcropping genes of slight or no selective value will develop, which would have had no chance of survival in a dense, stable population with high population pressure. The populations of *D. hydei* are actually very different in the various cities as far as type and number of mutant genes are concerned.

These effects of population size have been well described by S. Wright[20]:

"In a large population with sufficient random interbreeding and no secular changes in conditions over a long period of time, all genefrequencies may be expected to approach equilibrium values largely determined by selection. Once the population has reached a certain peak of adaptive values there will be no further significant evolutionary change in spite of continual mutation, persistent variability and rigorous selection. More rigorous selection will merely concentrate the species as a whole about the peak, raising the mean adaptive value, but reducing the variability on which further evolution must largely depend. The chance of occurrence of a wholly novel mutation, possibly adaptive from the first, is reduced, since the reduction in frequency of non-type alleles reduces the chance of occurrence of untried mutations at two or more removes from the type gene. Change of environment may depress the earlier adaptive peak, variation grows stronger and permits the development of a new adaptive peak, which better fits into the new conditions.

"Extreme restriction in numbers brings about a tendency towards random fixation of one or another combination of genes which is almost certainly less adaptive than the previous type. Many of them will be definitely deleterious. Whilst the variability of small populations is thus highly increased by "drift", the effect will certainly not be in the overwhelming majority of the cases in the direction of new adaptations or of progressive evolution. The combination of partial isolation of subgroups with intergroup selection seems to provide the most favourable condition for evolutionary advance."

This influence of population size on the variability of characters has been studied by FORD[21] in some Lepidoptera. He found that an isolated population of the Fritillary *Melittaea aurinia* in England has shown two prolonged peaks of abundance, within 49 years. Heavy "outbreaks of variation" occurred in the low population interval between the peaks, especially during the rapid increase after scarcity. Hardly two specimens were then alike in size, colour and pattern. During the periods of abundance, however, this high variability disappeared. A uniform type took its place. This uniform type was different for each of both periods of abundance. This observation strengthens ELTON's[22] view that strong periodic fluctuations have a marked evolutionary importance in animals. We can now enlarge the original topographical or geographical proposal of WRIGHT to time. The heavy contractions of populations serve exactly the same purpose: to permit a reshuffling of gene combinations which may lead to other and eventually more adaptive types.

We should like to add here one observation on the Vole *Microtus güntheri*, in which we occasionally noted a white gular spot in the field, yet only in years of beginning increase, never during one of the outbreak years.

Isolated populations show marked differences rather soon after the beginning of isolation. This has been mentioned before for populations of islands, of mountain valleys or of mountain peaks. Arctiidae and Zygaenidae, e.g. flying in different but neighbouring mountain valleys in the Alps or other high mountains are definitely in different pattern (PICTET, BOVEY). Another Fritillary, *Melittaea athalia*, was purposely introduced into a new, isolated area in England. This new Essex strain was very soon noticeably smaller and darker than the parent strain from Kent (STEVIN). Two local populations of the geometrid moth *Operabia autumnata* were recently only separated ecologically into the populations of two different types of woodland, yet they already show definite differences in size, colour and certain physiological characters (HARRISON).

An important problem has been raised by the establishment of melanistic forms of many moths in industrial areas. It was known for long that melanistic mutations crop out here and there in normal populations without, however, being able to maintain themselves. Suddenly, with the massive growth of industry, with its widespread deposits from fumes, the melanistic mutation grew in importance to form the large majority or even the total of the moth populations in these industrial areas, while undergoing no change in the agricultural areas. Obviously the melanistic mutation had suddenly gained vital superiority over the type form in the industrial areas. A film of the Oxford School of Genetics seems to show the great negative selective value which the melanistic form possesses in normal, the type form in industrial forests. A sudden liberation of equal numbers of moths of both forms showed a rapid preference of birds for that group which was unsuited to the environment. Has the scattered appearance of moths, in time and space, in nature, the same effect? We agree that obviously in one point the topotypic forms have positive selective value. We strongly suspect, however, that further physiological, selective differences in both forms support this one superiority. It is even possible that the melanistic phenocopies are stimulated by the chemical deposits on the leaves of the industrial areas.

Every progress in population genetics is welcome to the ecologists, as genetics in nature is precisely what is needed to bring the ecologist into reach of genetical analysis of ecological phenomena.

## 4. The Interaction of Environment and Heredity in the Diapause

A good illustration of the combined action of hereditary as well as environmental factors is the phenomenon of diapause. We shall here restrict the term diapause for any case of dormancy, interrupted development of maturation, which is not immediately broken by the change of unfavourable into favourable external conditions. Hibernation as displayed by the Mexican bean beetle, broken by contact humidity above a certain temperature-threshold, is here not regarded as diapause, whereas the hibernation of the eggs of *Bombyx mori* which is not broken before a certain interval is[22, 23, 24].

There exist a series of animal species which continue to develop uninterruptedly as long as they are under favourable conditions.

Besides many homoiothermous animals (voles, mice, etc.), flies like *Lucilia, Phormia, Calliphora,* gnats like *Phlebotomus*, wasps like *Hormoniella*, and many other insects belong to this group. COUSIN[25] has conclusively demonstrated that ROUBAUD's theory[26] of cyclic self-intoxication by urates after a sequence of generations, followed by an obligatory dormant period of purgation, is incorrect. Development is continued uninterruptedly as long as optimal conditions prevail, and diapause occurs as soon as one of the important environmental factors (food, temperature, humidity, population density, etc.) falls below a certain threshold of tolerance. The percentage of individuals entering diapause depends upon the age-distribution of the population (determining the number of individuals being in the sensitive age) and on the degree of unfavourableness of the factor or factors concerned. In this common case heredity has no part in the diapause, except in determining potential diapause as part of the normal reaction-base.

The reverse is true for another series of phenomena, rather common in the moderate climates of the Holarctic kingdom. Most birds and mammals with a seasonal rutting-period (e.g. goat, sheep, sparrow) and many insects with one annual generation only (*Bombyx, Lymantria, Cimbex,* etc.) belong here. Diapause sets in at a fixed stage of development independent of all possible environmental conditions, from optimal to almost intolerable, and is not broken even in favourable conditions until a certain period of time has passed. Under unfavourable conditions the diapause may last much longer, eventually until death sets in. The awakening from the diapause occurs in this case simultaneously in the laboratory and in nature. We have seen this often in *Cimbex humeralis* and *Eurytoma amygdali* in Israel[27]. In those cases hereditary diapause overrules all exogenous influences.

It is still entirely unknown as to how this timing of the diapause is regulated, especially as it almost always leads to a very "utile" synchronism between the optimal conditions of food, climate, etc., and the active period of the animal concerned. In certain cases, at least, the timing is not conditioned by time. GOLDSCHMIDT has shown for the egg of *Lymantria dispar*[28] and BODINE for those of *Melanoplus differentialis*[29] that the development inhibition of the egg is broken by low temperatures and that hatching depends on the sum of effective temperatures after the breaking down of the diapause. Winter cold is a regular phenomenon in the natural environment of these species, and GOLDSCHMIDT[28] is even inclined to conclude that every population (or race) of *Lymantria* is fitted genetically in the special climatic cycle of its natural habitat by small hereditary changes in the degree of cold necessary for breaking its diapause and in the value of its thermal constant.

Between those cases appearing to be wholly determined by heredity or environment occur many stages of transition in nature, where the coaction of both is very obvious. DAWSON[30] has shown for *Telea polyphemus*, BACOT for rat fleas[31], that a certain percentage of offspring grows dormant or develops parallel to the fall or rise of temperature during a certain sensitive period of previous development. This means that under certain conditions all pupae may become dormant. But under abnormal climatic conditions, e.g. with ascending temperature in autumn, some of the pupae, etc., go into dormancy anyhow, whereas others emerge under these favourable conditions. Recent research makes the duration of day responsible for many of these changes. This bimodal curve is decidedly genetically determined. But in the climate of Minnesota, favourable conditions in autumn lead the developing moths or their offspring to a premature death in winter. The obligatory partial hereditary diapause is therefore of vital importance to the maintenance of the species, whereas the partial facultative emergence would eventually open new developmental possibilities under other climatic conditions.

In some species, as in *Bombyx mori*, we find races with a different number of annual generations even in the same habitat. If more than one generation is present, one or two are of the emergent, one invariably of the dormant type. FOA attempted to correlate the vol-

**Table XXVIII.**

Constitutional types of pupal development of *D. euphorbiae* at 22° C

| Type | Designation | Development days | Loss of weight | Smallest $O_2$ consumption per $cm^3$. | Highest $PO_3$ % per mg | Reducing substances mg. % glucose |
|---|---|---|---|---|---|---|
| 1 | Biannual | 640 | Very slow | .. | Up to 70 | .. |
| 2 | Dormant | 270–330 | Slow | 20 | 50–55 | 50 |
| 3 | Lethal (dormant with final loss of weight) | Die during/after winter by excessive water loss | Quick | 20 | .. | .. |
| 4 | Protracted | 40–50 | Fairly quick | 35 | .. | .. |
| 5 | Emergent (=subitanous) | 17–21 | Quick | 50–60 | .. | 50 |
| 6 | Rapid | 13 | Very quick | .. | 30 | .. |
| 7 | Lethal soft winged | pupae die | .. | .. | .. | 220 |
| 8 | Lethal black pupae | Pupae die | .. | 160 | .. | 380 |

tinism of the silk-worm with the influence of ascending or descending temperatures on the early stages of the maturing egg, inducing the appearance of large or small cells in the germinal stripe[24]. Duration of day seems also here to participate.

Very complicated hereditary conditions affect the diapause of pupae of *Deilephila euphorbiae* near Lwow, as studied by HELLER[32].

Types 1, 2, 4, 5 are certainly genetically conditioned, 3 and 6 may be extremes of types 2 or 5 only. In types 7 and 8 it is difficult to ascertain whether lethal hereditary characters or individual cases of disease are the reason for the occasional appearance of these groups. Groups 2 and 5 predominate, as a rule, which also results in a bimodal curve. But more than one character is concerned in the determination of pupal diapause of *D. euphorbiae*, as is clearly demonstrated by the physiological analysis. Induction of emergence or dormany depends also in this case on environmental conditions (especially temperature) apart of type 2 and 1.

A complicated genetical problem is offered, where continued spontaneous development occurs side by side with obligatory diapause in the same population in the transition zone of the one- and two-generation area of a species, where the selection theory apparently requires the disappearance of the group with spontaneous development after a long period of negative selection[33].

Diapause is an organismic response to changes in environmental conditions, as a rule[34], but the difficulty of a non-ambiguous definition of diapause is obvious from the egg-development of *Dociostaurus maroccanus* in Iraq[35]. The embryo may arrest its development after the first formation of the germ-band for up to two months, whilst development proceeds slowly generally, without interruption. This is a facultative arrest of development. Then development proceeds extremely slowly for three to five months to a stage following anatrepsis.

This development is accomplished within one to three weeks at the same temperature in the emergent eggs of related species. This is a definite case of retarded or inhibited development. Finally, about September, when the stage between ana- and katatrepsis is reached, development ceases entirely, until a combination of sufficiently high temperature and water contact of the soil – which until that stage had had no accelerating effect – breaks the diapause, normally at the end of winter. Then final development and hatching take place within a few weeks.

That phase has always been called diapause. It definitely is arrested development, which is, however, interrupted immediately with the appearance of the required combination of favourable exterostimulations, i.e. it is really a readiness for immediate development under proper conditions, which in nature do not occur during winter.

In later stages of development (larva to adult) increasing material

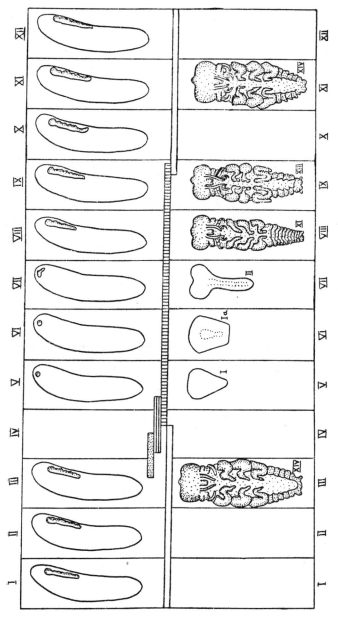

Fig. 36. The egg-development of *Dociostaurus maroccanus*. Month by month, in its natural environment in N. Iraq, demonstrating different types of diapause. Above: Size and position of the embryo in the egg. Centre: Sequence of normal seasonal stages: dotted (larva), horizontal lines (adult), vertical lines (eggs of phase A and B), empty rectangle (eggs of phase C and D). Below: Stage of embryo most common in every month (From BODENHEIMER & SHULOV, 1951).

is available of hormonal and neurosecretory regulation of all processes connected with diapause since the pioneer work of WIGGLESWORTH, JOLY and WILLIAMS. We have nonetheless hesitated to apply similar conceptions to the egg, although this is wholly unfounded as the enzyme mechanisms of the early development of the egg are actually comparable to hormonal or neurosecretory regulations, while in the later stages of embryonal development incretory glands and nervous systems are already available.

The varying life-history of the polyphagous egg-parasite, *Trichogramma*, its dependence on the special life-history of its special host, its seasonal cycle and diapause, have been submitted to a masterly analysis by MARCHAL[36].

Similar phenomena are by no means unknown in mammals and birds. The seasonal fixation of sexual activity and of the coincident hormonal secretions are one of them. Fig. 37 shows how deeply fixed the seasonal rhythm of the natural habitat, in a group of nearly related gazelles, is in agreement with environment, mainly rains and vegetative period. In the Cairo Zoo where green food supply and climate are even throughout the year and where no rainy season occurs at all, this rhythm of the original habitat persists with regard to births[37].

## 5. The Interaction of Environment and Heredity in Sex-Determination

Sex-determination yields another splendid illustration of coaction between genetical and ecological factors.

The discovery of the sex-determining X and Y chromosomes in every living plant and animal by McCLUNG, WILSON, MONTGOMERY, and others, at the beginning of this century, and its confirmation by the mono-sexualism of the offspring of the one egg-polyembryonism in the armadillo *Tatusia*[38], the chalcidid wasps *Encyrtus* and *Ageniaspis*, the different chromosome structure in both body halves in bilateral gynandromorphs, etc. seemed to furnish a decisive and definite defeat to any theory considering any environmental influence whatsoever on sex-determination[39, 40].

However, whether we like it or not, there is no doubt that quite a series of well-known facts are not easily explained by the orthodox chromosome theory. One of the most convincing is the fact that most organisms still seem to contain the potentialities of both sexes. This is certainly true for habitual hermaphrodites such as most earthworms, snails, and tapeworms; for cyclic hermaphrodites, producing alternately sperms and ova as in *Ostrea edulis*[41], or in successive hermaphrodites producing first sperms and later on eggs only as in *Icerya purchasi*[42]. STEINACH's experiments on feminization or masculinization of the opposite sex in secondary sexual

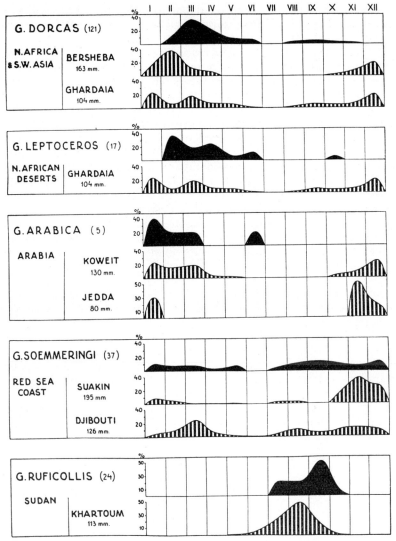

Fig. 37. Correlation between the monthly percentage of births of some gazelles (black) in the Cairo Zoo (with optimal and equal feeding throughout the year – as reported by FLOWER, 1932) with the rainfall in their natural distribution areas (striped).

characters, as well as in behaviour, by castration followed by transplantation of the opposite sex glands, are a further proof of these potentialities[43].

Most convincing are the experiments with poultry or pigeons.

When the developed ovary of a hen, the left one, is operatively removed, the right ovary, suppressed and undeveloped hitherto, begins to swell. It then develops not into an ovary but into a normally functioning testis, producing sperms as well as internal secretion. In consequence, the whole exterior of the hen, as far as it is based on secondary sexual characters, changes into that of a cock. Even the uterus and vagina may change to such a degree that these individuals are able successfully to fertilize normal hens[44].

In the amphibians Bidder's organ probably always develops into a half-functioning ovary. If the testes are extirpated in the males of *Bufo viridis*, the individuals become normal ovipositing females. The lability of amphibian sex-glands during the early development is a long and well-established fact. WITSCHI obtained from tadpoles of *Rana silvatica*, bred at high temperatures, or from eggs which were over-matured experimentally, male frogs only[45].

Most impressive are the experiments of BALTZER and others with the marine worm *Bonellia viridis*[46, 47]. Male larvae grow into females, if they have no opportunity to settle on and suck the pharynx of an adult female. If female larvae settle on and suck the pharynx of an adult female they change into males. Chemicals, such as weak solutions of coppersulphate, also have a sex-determining effect. Intersexes developing from female larvae which have not sucked sufficiently long on the pharynx of an adult female are known. After a certain period of attachment all these larvae grow into males.

We learn that in *Daphnia* environmental conditions during a critical period of egg-development determine the chromosome character and therewith the sex[48, 49, 50]. Among these environmental conditions the size of the available life space has an important influence. This fact is confirmed in other groups, e.g. in Ichneumonidae or in Mermithid worms:

**Table XXIX.**

Sex of *Paramermis contorta* in chiromonid larvae

| No. of parents per larva | No. of females | No. of males | Ratio of female: male |
|---|---|---|---|
| 1 | 255 | 17 | 15.00 |
| 2 | 120 | 106 | 1.14 |
| 3 | 47 | 82 | 0.57 |
| 4 | 23 | 41 | 0.56 |
| 5 | 5 | 25 | 0.20 |
| 6 | 3 | 15 | 0.20 |
| 7 | 3 | 11 | 0.27 |
| 9 | 1 | 8 | 0.13 |
| 10 | 4 | 26 | 0.15 |
| 11 | 2 | 9 | 0.22 |
| 17 | 2 | 15 | 0.13 |

Population pressure – in this case possibly either quantity of food or of a sex hormone available per individual – leads to strong prevalence of the male sex.

This type of sex-determination leads to another ecologically important case. Many animals reproduce parthenogenetically for a series of generations, producing sexual generations during certain seasons. WEISMANN and others attempted to explain these changes, which occur in aphids, Cladocera, and rotifers, as genetically induced. Observations extending over long years, combined with laboratory experiments, showed that under optimal conditions these animals may reproduce by parthenogenesis indefinitely and that the appearance of the sexual part of the cycle is always induced by unfavourable environmental conditions. The work of LUNTZ[51] and WESENBERG-LUND[52] on rotifers, the research on Cladocera[53], the behaviour of aphids[54] in tropical and subtropical countries, prove these descriptions to be true, as do also the recent investigations of HARTMANN, KLEBS, etc. on freshwater protists, in which unfavourable environmental factors were confirmed to induce depression connected with conjugation.

Environmental factors induce the appearance of males in the sexual cycles, environmental factors (temperature) induce the alternating sperm and egg-producing cycle of the oyster. Environmental factors induce the male sex in frogs (temperature, maturity of egg), in *Bonellia* (chemical stimuli), in Mermithids (density). All these facts are not explained by differential mortality of both sexes in early stages of development, an explanation which is often and probably correctly used for the unequal sex-ratio in man. All facts recorded here are incompatible with the conception of the sexes as two different qualities of the living matter definitely determined by chromosome pattern. They are intelligible when sex is considered as a form of polar differentiation of the same quality. And all modern theories which consider the intersexes too, start from this point of view.

There exist still purely morphological theories, as that of BRIDGES[55], which tries to explain the sex as a numerical relation between X-chromosomes and autosomes in *Drosophila:*

| Sex-condition | No. of X-chromosomes | No. of autosomes | Ratio autosomes/ X-chromosomes |
|---|---|---|---|
| Over-females . . . . | 3 | 2 | 0.67 |
| Females . . . . . . | 4 | 4 | 1.00 |
| Females . . . . . . | 3 | 3 | 1.00 |
| Females . . . . . . | 2 | 2 | 1.00 |
| Intersex . . . . . | 2 | 3 | 1.50 |
| Male . . . . . . . . | 1 | 2 | 2.00 |
| Over-male . . . . . | 1 | 3 | 3.00 |

In the physiological theory of sex-determination of GOLDSCHMIDT[56], Mendelian intersexes occur regularly when crossings are made between certain races of *Lymantria dispar*:

| Crossings | | | Results |
|---|---|---|---|
| ♀ Hokkaido (weak) | × Hokkaido (weak) | ♂ | Only purely monosexual female offspring. |
| | × European (less weak) | ♂ | Only purely monosexual offspring, but eventually weak intersexes. |
| | × Gifu (medium) | ♂ | Heavy male intersexes. |
| | × Ogi (strong) | ♂ | Practically only offspring of male phenotype. |

We have here a similar quantitative series without correlated changes in chromosome ratios. GOLDSCHMIDT assumes that male and female potentialities are present in every body, both being present in certain intensities (♂ 120, 100, 80; ♀ 100, 80, 60). If the difference of both intensities is at least 20 units, pure sexes are produced, whereas slighter differences induce the appearance of intersexes. The beginning of the reaction determines the degree of intersexualism. GOLDSCHMIDT conceives, therefore, that the sex-genes are connected with intensity differences.

RIDDLE[57] leaves the chromosome base of sex-determination still farther behind, in stating that in eggs of frog, toad, or pigeon all conditions which increase the oxidation rate favour the tendency towards male, all those which decrease it favour the female tendency. In vertebrates the male sex has constantly a higher intensity of metabolism (5–14%). The lipoid content of blood is lower in males than in females, therefore metabolism is higher in the former, as is known for microgametes in Protozoa. For locusts and some other insects the author can confirm that metabolism is always higher in the male sex (related to the weight-unit). This would mean that the X-chromosome does not determine the sex, but the general intensity of basic metabolism, which in turn determines the sex. This agrees with studies on metabolism of the oyster, which is quantitatively different in both sexual phases. And we have no difficulty in understanding that basic metabolic intensity may be changed by environmental influences. This change of metabolic intensity leads to a physiological as well as structural and behaviouristic reconstruction of the body, accompanied by change of sex. We can thus conclude, with regard to the problem of sex-determination that it provides a good illustration of the coaction of heredity and environment within the organism.

## 6. The Interaction of Environment and Heredity on Geographical Variation

As far back as 1926 we stressed[58] the fact that the factors determining the fluctuations of a species in time are the same as those which determine its spread and distribution in space. At that time an outbreak of *Chaerocampa celerio*, so far unparalleled in this country, took place in the vineyards under weather condition which agree with the climate of some areas in South Africa where the species causes regular damage. Thus, epidemiology and zoogeography are complementary resultants of the same groups of factors. This view has since been widely accepted, as may be seen, e.g. from the title of the recent book "The Distribution and Abundance of Animals" by ANDREWARTHA & BIRCH[59].

Geographical variation is the change of certain characters, transitory as a rule, within one species, with the shift of its areas within its range of distribution. Whereas a few species of a very large area, as the Painted Lady *(Pyrameis cardui)*, show almost no variation, in most animals changes occur in individuals from the eastern or northern end of their area to the western or southern border. The transition is gradual and the variation ranges of two neighbouring areas overlap often. This holds for changes in size, colour, morphology of certain organs, etc. Fig. 38 illustrates this point.

A further very common trait of geographical variation is the simultaneous shift in a series of characters.

The large woodpecker *(Dryobates major)* is represented by two forms in central Europe[59]; the western form *D.m. pinetorum* and the eastern race *D.m. major*. These differ in the following characters:

| Character | D.m.major | D.m. pinetorum |
|---|---|---|
| 1. Length of wing | 136–145 mm. | 131–138 mm. |
| 2. Length and width of eggs | 26.9 × 20.7 | 25.7 × 19.3 |
| 3. Ratio wing length/ beak length | 20.6 | 22.0 |
| 4. Form of beak | Slender, steadily becoming narrower | Remaining thick, broad from base to middle of beak |
| 5. Colour of lower side | Mainly white | Mainly light brown |

Five characters, at least four of which are heterogeneous, show a simultaneous and steady alteration with territorial shift. Extended correlated alteration of morphological and physiological character has been described by GOLDSCHMIDT[61] for the caterpillars of *Lymantria dispar*.

This modification of characters is certainly not a uniform phenomenon from the genetical point of view. There are characters which seem to depend mainly upon the intensity of certain environmental factors, and there are others which most probably are independent of such influences.

The first group often leads to phenomena commonly regarded as adaptive. The few fundamental general rules of zoogeographical variation belong to this group.

BERGMANN's[62] rule states that races of mammals and birds living in different climates increase slowly in size towards the colder habitats of their area. Cold delays sexual maturity, which is – also relatively – accelerated by high temperature. The period of growth is lengthened. The cell size remains constant, but the number of body-cells increases in consequence of this prolonged period of growth. This simultaneously induces a smaller surface/volume ratio in colder climates, economizing thus the loss of heat by radiation.

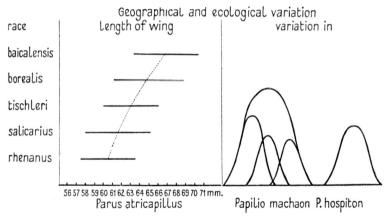

Fig. 38. Geographical variation. A. Range of variation in wing length of geographical races of *Parus atricapillus* (in mm). The dotted line connects the average values (after RENSCH). B. Ecological and geographical variation in the Swallowtail *Papilio machaon* as compared with the variation of the same character in the closely related *Papilio hospiton* (after ELLER)[60].

The same influence of temperature on the growth is often met with in poikilothermous animals (snails, amoebae, etc.), but any other factor may interfere with this tendency. Unfavourable conditions of humidity, of food, lime content of soil, etc., immediately reduce the size, thus interfering with the mere temperature influence. Many positive observations on this have been compiled by HESSE[63] and RENSCH[59].

ALLEN's rule: Under different climatic conditions races of mammals and birds show a considerable reduction in the relative and often in the absolute size of body projections, such as tail, ears, and extremities, towards the colder habitats[64].

In the American woodpecker *Dryobates villosus* the relative wing length[59] is reduced in the north.

| Race | Area | Total length | Wing length | Revative wing length |
|---|---|---|---|---|
| D.v. septentrionalis | Canada, Montana | 248 | 132 | 53 |
| D.v. leucothorectis | Arizona, New Mex. | 216 | 126 | 58 |
| D.v. extimus | Costa Rica, Panama | 172 | 103 | 60 |

The reduction of ear-size in hares towards the north has been described by HESSE[63] and others. It is probable that exposed body parts suffer from growth inhibition, reduction of vascular system, etc., as a direct consequence of exposure to cold. Under the influence of cold, these thin and exposed parts anyhow show a temperature far below that of the body. A few examples exist also with regard to insects. But the mouth parts of the honey-bee grow smaller in warmer climates[65]:

| *District* | *Tongue length* |
|---|---|
| Orel district (Central Russia) | 6.730 mm. |
| Ukraine | 6.549 mm. |
| Italy | 6.234 mm. |
| Israel (Jerusalem) | 6.292 mm. |

GLOGER's[66, 59] rule: The melanins of mammals, birds, and of perhaps all animals intensify with increasing temperature and air humidity. Heavy melanin accumulation in bird feathers leads to the appearance of interference colours. Dry, hot weather increases the phaeomelanins only, leading to yellow desert-coloured degrees of oxidation, whereas the eumelanins disappear. Strong cold inhibits first the formation of the phaeo-, later of the eumelanins resulting finally in the "white" polar coloration[67]. The thorough knowledge of coloration in one group may enable a fair climatic description of the natural habitat of any race. HINTON[68] says with regard to the African squirrel *Heliosciurus rufobrachium:* "When one knows how to read the characters of this squirrel one can describe the physical condition of the locality from which any given specimen came without knowing either the name or the whereabouts of the locality."

Many other rules can easily be formulated. But it seems inadvisable to do so without a thorough revision of the relative facts. Several of such rules state:

Fur and feathers of mammals and birds grow thicker in colder climates.

Tropical birds lay fewer eggs than the same species in a moderate climate (perhaps owing to the shorter length of day in the tropics)[59].

The relative heart-weight of birds and mammals increases in the same species with increasingly colder climate, etc.[63].

The characters concerned in these three rules are all hereditary. But convergent phenogenetic characters may be produced in the experiment. SUMNER[69] bred genetically equivalent lots of mice at different temperatures with the following results (after 2 months):

| Temperature | Weight | Length of | | | Weight of hair of fur |
|---|---|---|---|---|---|
| | | Tail | Ear | Hind foot | |
| 26.3° C | 17.0 g | 93.2 mm. | 16. mm. | 21.4 mm | 264.6 mg |
| 6.2 | 17.9 | 75.9 | 15.9 | 20.9 | 294.8 mg |

The same holds good for insects, as was demonstrated by the classical experiments of STANDFUSS[70], who obtained in Switzerland from local pupae of *Vanessa antiopa*, kept in cold, the Arctic form of this species; and from local pupae of the Swallow-tail *Papilio machaon*, kept at high temperatures, the Israel form of this butterfly.

In discussing the importance of the "dauermodifications" for hereditary fixation, the time factor should not be overlooked. If a species may "remember", so to speak, for some time after they have ceased to act, external stimuli, which, as in the experiment, have worked for a few generations only, how much deeper should be the fixation of this memory after the stimulus has worked for secular or even minor geological periods?

It does seem logical to assume that many of these hereditary geographical races originated as phenovariants, which became genetically fixed under the constant and equal influence of climatic factors through many generations[59].

Connected with this conception is the dauermodification of JOLLOS[71]. Under the influence of environmental conditions certain characters of an organism as well as of a population change. This change is reinforced or retained as long as the inducing external conditions continue. When they cease and the normal environment is restored, the change of character is retained for some period and will continue to appear in a gradually decreasing degree still for some generations until it finally disappears.

**Table XXX.**

Induction of half-moon formation in *Arcella polypora* by selection

| No. of selection | Date | A Under normal cultivation | | | | | B At environment favouring halfmoon formation | | | | | Remarks |
|---|---|---|---|---|---|---|---|---|---|---|---|---|
| | | 0 | 1 | 2 | 3 | 4 | 0 | 1 | 2 | 3 | 4 | |
| 0 | 30.X.22 | .. | 1 | .. | .. | .. | .. | .. | .. | .. | .. | Selection of plus variants. A. Under normal cultivation. B. Under conditions favouring half-moonformation. The nos. from 0 to 4 describe the different stages of half-moon formation from normal (0) to complete (4). |
| 1 | 8.XI.22 | 75 | 219 | 6 | .. | .. | 22 | 203 | 56 | 19 | .. | |
| 2 | 16.XI.22 | 24 | 267 | 7 | 2 | .. | 5 | 107 | 101 | 87 | .. | |
| 3 | 27.XI.22 | 5 | 207 | 59 | 29 | .. | .. | 9 | 98 | 164 | 29 | |
| 4 | 5.XII.22 | .. | 82 | 101 | 117 | .. | .. | 2 | 7 | 86 | 205 | |
| 5 | 13.XII.22 | .. | 34 | 113 | 144 | 9 | .. | .. | .. | 12 | 288 | |
| 6 | 22.XII.22 | 1 | 24 | 68 | 186 | 21 | .. | .. | .. | 3 | 297 | |
| 7 | 3.I.23 | .. | 10 | 7 | 75 | 208 | .. | .. | .. | .. | 300 | |
| 8 | 15.I.23 | 6 | 11 | .. | 80 | 203 | .. | .. | .. | .. | .. | |
| 9 | 25.I.23 | .. | .. | 2 | 43 | 255 | .. | .. | .. | .. | .. | |
| 10 | 5.II.23 | .. | .. | .. | 21 | 279 | .. | .. | .. | .. | .. | |
| 11 | 25.II.23 | .. | .. | .. | 8 | 292 | .. | .. | .. | .. | .. | |
| 12 | 1.XII.23 | .. | .. | .. | .. | 300 | .. | .. | .. | .. | .. | |
| | 5.VII.23 | .. | .. | .. | .. | 1 | .. | .. | .. | .. | .. | Beginning of selection: breeding in order to obtain minus variants in normal cultivation. |
| | 15.VII.23 | .. | .. | .. | .. | 300 | .. | .. | .. | .. | .. | |
| | 24.VII.23 | .. | .. | .. | .. | 300 | .. | .. | .. | .. | .. | |
| | 31.VII.23 | .. | .. | .. | .. | 300 | .. | .. | .. | .. | .. | |
| | 12.IX.23 | .. | .. | .. | 14 | 286 | .. | .. | .. | 13 | 287 | |
| I. | 22.IX.23 | .. | .. | 8 | 59 | 233 | at normal environment | | | | | Breeding by selection of minus variants in normal cultivation. (B. without selection.) |
| II. | 13.X.23 | .. | 26 | 95 | 139 | 40 | | | | | | |
| III. | 20.X.23 | 1 | 213 | 66 | 15 | 5 | .. | 82 | 32 | 76 | 110 | |
| IV. | 27.X.23 | 24 | 262 | 14 | .. | .. | .. | .. | .. | .. | .. | |
| V. | 5.XI.23 | 259 | 41 | .. | .. | .. | .. | .. | .. | .. | .. | |
| VI. | 12.XI.23 | 300 | .. | .. | .. | .. | 113 | 109 | 16 | 47 | 15 | |
| | 15.XII.23 | .. | .. | .. | .. | .. | 210 | 57 | 12 | 11 | 9 | |
| | 14.I.24 | .. | .. | .. | .. | .. | 272 | 16 | 4 | 6 | 2 | |

In ciliates similar dauermodifications have been obtained for toxin-tolerance. In some cases the effect of the dauermodification disappears abruptly after the first copulation in normal conditions, after which act the nuclear genotypic mechanism subdues the plasmatic influence. SONNEBORN and others have discovered plasmogenes as additional bearers of heredity in ciliates. In other cases, however, the modification is still maintained in some strength even after half a year and after many copulations.

The most plausible explanation is offered for dauermodifications in metazoa, such as have been observed in crustaceans and in insects. We[72] have made some experiments on the influence of some substances in various concentrations on the length of development in *Drosophila*. For every substance we found a concentration at which development is – without any specific effect – significantly reduced. The normal development from egg to emergence was 190 hours. In the following series the breedings were at optimal concentration of these substances and salts. The difference ($m_{diff}$) between $F_1$ and $F_4$ was in all cases 3 to 10 times the limit of significance (3). The variation of development was low.

Table XXXI.

Reduction of development in *Drosophila* by various substances

| Substance | Duration of development in hours | | |
|---|---|---|---|
| | $F_1$ | $F_4$ | $F_5$ |
| Thyroid | 189 | 169 | 172 |
| Thymus | 190 | 173 | 170 |
| $NaNO_3$ | 190 | 170 | 171 |
| $CuSO_4$ | 189 | 169 | 170 |
| KCl | 189 | 167 | 169 |

After the return of the breedings to normal agar, it took five generations or more to return to the duration of development of the original parent generation. Yet, as in all other cases of dauermodification in metazoans known, this is only true for the females, which when crossed with normal males, retained some definite acceleration of development. The offspring of males, however, when crossed with normal females, returned immediately to the speed of development of the parents. The explanation appears obvious: In the big cytoplasma mass of the egg, a progressive thinning concentration of the inductive substance is maintained which grows thinner, i.e. weaker, from generation to generation, until it disappears. In the almost negligibly small cytoplasma mass of the spermatozoa this influence is abruptly lost in the big cytoplasma mass of the normal egg. This assumption could be put to proof easily if tracer molecules of a longevity of some months – which were not at our disposal here – could be added to the substances. The role and character of the plasmogenes of some ciliates also requires further elucidation.

There are other characters, as the genitalia of most insects, which show rich geographical variability without recognizable parallelism between any environmental factor and variation tendencies.

We are far from understanding conclusively how the genes within the chromosomes, mutations, dauermodifications, and environmental influences coact. There is little doubt that most of what we call geographical variation is some type of dauermodification within the range of a specific reaction basis or constitution. In *Papilio machaon*[60], in *Lymantria dispar*[61, 73], and in some other cases this has been demonstrated clearly not to be the case. But these negative results should not be understood to exclude geographic variation in all cases and to be the only way of species formation. Cases of transition of monophagous insects to new hosts and subsequent isolation of two biological races (*Carpocapsa pomonella* and *Rhagoletis pomonella* in North America) which have not yet been analysed by the geneticist, are possibly beginnings of species formation[74].

We now leave this topic without trying to understand the way of formation of new species. The material at hand is still utterly inadequate to approach an adequate solution. But we would like to point out that different ways, and not one, are realized for this purpose in nature. Sudden high mutability induced by extrinsic as well as by intrinsic factors, possibly sometimes directed, but certainly more often quite undirected; permanent plasmotic fixation of phenotypical reactions of which the dauermodification seems to be a first step – these and still other ways are all probably verified occasionally.

## 7. The Problem of Adaptation

Finally, we shall deal with the problem of adaptation. Some of the rules of geographical variation, expressing direct parallelism between tendency and intensity of variation as induced by certain environmental factors, are regarded as adaptive ones. In an earlier stage of biological development these were considered the result of selection among mutants. It has, however, been repeatedly pointed out that most of the animals and birds of the desert showing extreme desert colouring are nocturnal, i.e. would not benefit at all from the "protective" colouring, whereas some of the diurnal species of the same habitat are black, i.e. very conspicuous.

HEIM DE BALSAC[75] has recently shown that the hypertrophic development in the auditory bulla in desert mammals is a common result of the influence of desert conditions, and that no plausible hypothesis as to its physiological importance could be suggested*. In polar animals the white colour might be of selective importance and certainly is so occasionally. But the tendency to white colouring by inhibition of melanin development in cold remains a primary fact. The same is true for the wingless insects of all classes in the Kerguelen Islands[76]. The strong winds are certainly unfavourable to the maintenance of weak fliers with large appendages, i.e. with great surface resistance. However, cold induces short-wingedness by inhibition of tyrosinase development, as has been shown experimentally by DEWITZ[77]. And results of field- as well as laboratory research in plants lead to the same conclusions:

"Of approximately a hundred species employed in adaptation studies, none has failed to respond to intensity or duration of factor stimuli, and so uniformly in nature and extent as to leave little doubt that adaptation in plants at least is a universal process, and not a selection of genetic strains or variations. This is indicated likewise

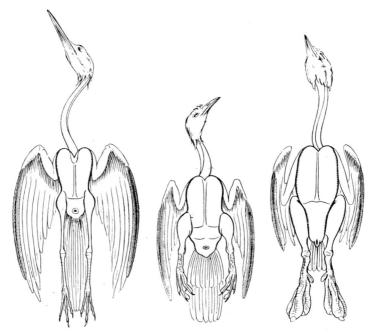

Fig. 39. Ventral view of 1. *Eurypyga helios*, 2. *Heliornis fulica*, 3. *Podilymbus podiceps* from Brazil. For explanation see the text (after BOEKER, 1930).

---

* F. PETTER (1956) has recently proposed a utilitarian theory for micromammals living at low population density and thinks that the phenomen is only restricted to desert animals living under those conditions.

by the different types of material, no species having yet been discovered in which the clones differ from the adult individuals in the response to the various factors[78]."

The agreement observed between form and function is the main reason why many palaeontologists and comparative anatomists are inclined to describe some kind of heredity of acquired characters as a "postulate of practical reason". The recent research of BOEKER has shown in many cases the extent of the morphological reconstruction of the whole body when a species leaves the normal habitat of its group and lives in a very different environment.

The Rallid bird *Heliornis fulica* of Brazil left the habitat of its nearest relatives in swampy primeval forest and became aquatic, adopting diving habits. Fig. 39 shows how this type of living has influenced the construction of the body. Relics of the land habitat are seen in the long neck and head, the long feathers on wing and tail, the strong development of breast musculature. Adaptations to the water habitat are the lobed toes, the short legs, the strong thigh musculature, the broad hip, the rich fat-deposit in the skin, the large size of the rump gland. Some measurements, related to a size of 20 mm body-length are[79]:

| Habitat | Forest swamps | Swimming and diving in water | |
|---|---|---|---|
| Family | Rallidae | Rallidae | Podicepidae |
| Genus | *Eurypyga* | *Heliornis* | *Podilymbus* |
| Length of shank, mm | 18 | 7.5 | 4.5 |
| Length of rump gland, mm | 2.6 | 7.5 | 6.7 |

The change in manner of life has obviously caused a change of body construction which gives to the Rallid *Heliornis* the appearance, at first sight, of a true Podicepid.

In studying certain organs of animals we find a surprising coincidence between the form and construction of this organ and the general mode of living of the species.

DOR has recently studied the morphology of the mammal tail in connexion with locomotion. The tail is a splendid object for such studies as it is not a vital organ as regards internal physiology and shows a high degree of variability with regard to osteology, myology, integument, and general form. The degree of relation between development, form, and function of the tail and the mode of living of mammals is surprising. DOR has never met with two species of the same habitat and type of locomotion which were principally different with regard to tail structure. On the contrary, many cases may be quoted where related species had very different tails when living in different habitats or having a different type of locomotion.

"The changes of the morphology of the caudal appendage are due to – sometimes delicate – modifications in the animal's way of living. Evidently the caudal appendage, having no determined and constant role, cannot be considered as a factor able to enforce, on its part, the mode of living. On the contrary, it is the tail which adapts itself to the biological conditions of the animal, especially to the mode of locomotion, the main expression of its activity[80]."

HESSE in his classical *Tiergeographie auf ökologischer Grundlage* describes an overwhelming number of morphological changes in animals which occur parallel to changes in certain environmental factors. The form of cross-section in fish living on the bottom (flat in dorse-ventral direction), in quiet water (compressed laterally), and in running water (round) may be quoted. Convergent development of certain characters as reduction of eyes, lengthening of extremities and antennae, loss of pigmentation in cave and subterranean animals is another illustration.

It is, however, unwise to exaggerate the value of these convergencies. RABAUD[81] points out that the morphological "adaptations" are by no means as ideal as is often claimed. Palmated feet in waterbirds are by no means in direct parallelism to swimming power. Whereas the slow and poorly swimming swan has very complete palmate feet, there are only small lobes on the feet of the coot *(Fulica atra)* which is a quick and strong swimmer. *Rallus aquaticus* and *Gallinula chlorops* have free toes, but swim well, whereas many ducks with well-developed webs are poor swimmers. Furthermore, palmated feet are developed in many terrestrial birds of different families, *(Columba palumbus, Phasianus colchicus, Caccabis saxatilis, Camprimulgus europaeus)*. This means that from the point of view of the palaeontologist or comparative anatomist not all organs are liable to be reconstructed to an equal degree.

Another series of facts deserves our attention, such as the formation of new races by change of host or by host-specialization. Only a few experiments and observations will be quoted.

Plant-feeding Nematodes prefer, as a rule, to feed upon the same host species or even variety upon which their parents have lived. This preference increases with the number of generations a population lives on it, the Nematode thus getting more and more specialized on this particular host. The injury inflicted to the host from a specialized strain is correspondingly more serious than that caused by a generalized strain. "Finally the specialization may reach such a high degree that even new hosts of the closest taxonomical, physiological, and chemical relationship to the old host are attacked no more or very lightly[83]." The probability or successive selection among mixed Nematode populations has become rather small with the progress of experimentation. Fig. 40 illustrates a rather interesting phase concerning one of the oldest experiments with plant-nematodes.

Man has witnessed the invasion of walnut in California by the codling moth *(Carpocapsa pomonella)*, both races, that living on walnut and that on apple, being now apparently unable to develop on the respective host of the other variety[84]. Forced transition to new host has been repeatedly observed. *Stilpnotia salicis* in Canada had to leave by force its original European host *Populus italica* for the Canadian *P. trichocarpa*[85]. The mortality of young caterpillars was heavy in the 1st generations, but normal later on. When retransfer to the original host was tried experimentally the same heavy mortality occurred during the 1st generations.

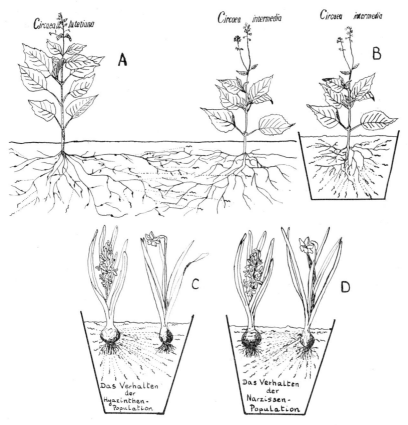

Fig. 40. Host selection of plant nematodes. A. at choice between *Circaea lutetiana* and *C. intermedia*. B. on *C. intermedia* as a forced host. C. of the hyacinth strain when hyacinths and narcisses are offered. D. of the narciss strain, when narcisses and hyacinths are offered (after STEINER).

Harrison[86] produced experimentally a new biological race of *Pontania salicis* adapted to *Salix rubra* from a pre-existing *S. andersoniana* race. Here also initial mortality of the young larvae was high, but after a period of 4 years a flourishing colony was established on the new host. Specimens of *S. andersoniana* were now planted among the other willows, but during 3 subsequent years of observations none of the sawflies showed an inclination towards returning to this plant. Harrison concludes that the modified egg-laying preference had become germinally fixed.

That oligophagous insects often – but certainly not always – tend to prefer for oviposition the host in which they have self-developed is a well-established fact, known as "the host selection principle"[87]. Such preferences are not always exclusive or alternative ones. Quite commonly only a certain tendency is observed, as in the case of two races of *Yponomeuta padella* observed by Thorpe[82]. Females of both races were given equal opportunity to oviposit on hawthorn and blackthorn and on apples:

| Moths | Eggs and percentage of total oviposition deposited on | |
|---|---|---|
| | Hawthorn | Apple |
| Hawthorn race | 911 (79.3%) | 237 (20.7%) |
| Apple race | 367 ( 9.8%) | 3,397 (90.2%) |

The number of eggs was about equal in all ovipositions. Each race preferred its own normal food, but would have taken freely to the alternative host if forced to do so.

After a full discussion of all known cases of biological races in insects and allied groups, Thorpe[82] comes to the following conclusions:

"We may say that many of these experiments are easily explained on some form of Lamarckian theory, but extremely difficult to account for on any other lines. If the hypothesis were not such a debatable one the evidence might well be regarded as almost conclusive. It seems equally certain, however, that none of the experiments has been on a sufficiently extensive scale to carry complete conviction on the point ... Taken together, they provide a quite considerable amount of the ever-growing body of circumstantial evidence for the theory."

Later Thorpe discovered the great influence which the maternal odour at the time of the emergency of the adults has on the place of their oviposition[88].

A striking case among the few others in which a real adaptation into a new environment seems debatable is the experiment of

DALLINGER[89]. This American student worked with three species of Flagellata which first showed an upper temperature limit of 23° C. When the temperature approached the upper limit, heavy vacuolization of the protoplasm occurred, preceding death. But when the culture was kept at 21° C for 1–3 months, it could be shifted to 23° C, then after the same interval to 25° C, and so on, until, after 7 years, the species lived at 70° C, a temperature not endured by any animal in nature. The experiment broke down due to disorder in the thermostat[90].

There are only two possible explanations. The sojourn within the high temperature may induce a physiological reconstruction of the organism which induces the shift of its upper heat tolerance. Cases of such reconstructions are well known and the case of *Aurelia aurita* from Nova Scotia and from Florida may be cited[91]. The orthodox geneticist denies this possibility, believing that the heat induces many mutations, some of which show a higher heat tolerance than the type population. What occurs after the shift of temperature is negative selection, only the new mutations fitting into the changed environment being able to survive. Such cases in which previous chance mutations enable the organism (but not its original population) to enter a new environment are called preadaptations[92]. This controversy between real adaptation and preadaptation regularly re-occurs in any case of pretended adaptation, and it is most unfortunate that no experimentum crucis has yet been discovered to decide between either proposed solutions. The answer does not need to be an alternative one, but possibly, or even probably, both cases are realized in nature. Environmental conditions often induce a physiological or morphological reconstruction of the organism. These are active responses of the animal to external stimuli and no mere physical responses to physical actions. Those reconstructions are either at first phenotypical, and may become later dauermodifications, or they are based on mutations (pheno-copies).

All these responses of the organism are primarily without any relation to utility, being merely responses to stimuli. In some cases, as in those of the adaptation of animals to the local climatic cycle, in the white colouring of polar animals, in the surface decrease of larger animals in colder climates, etc., the animal decidedly benefits from those responses. In many or perhaps even most cases no benefit at all is recognizable. In no case does this response surpass the hereditary base of reactions, and therefore no real adaptation to a new environment ever occurs. What is called adaptation is really a response to environmental stimuli within the hereditary reaction base of the species, but without preconsideration of the utility of the response towards the changed environment. Whenever only the responses fitting well into the new environment are understood as adaptations, this should be specially stated.

Still another series of facts, far from proving Lamarckianism, is difficult to explain on another basis, such as the thicker skin of the sole of the human embryo in contrast to the dorsal skin of the foot (SEMON); the peculiar shape of certain foot bones (astragalus) in the embryo of people accustomed to squatting in certain positions conforming with the shape which would have been achieved in the born individuals through usage; the horny callosities in the embryos of wart-hogs *(Phacochaerus)* and of camels in those places where the adults kneel and where such callosities would mechanically be produced, etc. We insist on mentioning the peculiar cases of marsupials' hair direction in fur, which conform in the embryo to that established by the toilet habits of the adults (WOOD-JONES[94]). In the elongate body of the long-legged kangaroo *Wallabia greyi* the hairs slant from the head towards the tail, except on a dorsal and lateral area, where the legs stroke the hair headwards during the toilet – the embryo has the hair in this area already slanted headwards, in contrast to the rest of the body. In the short-bodied long-armed Koala *(Phascolarctus cinereus)* the body hairs are slanted headwards, quite unusual in marsupials, except for a medio-dorsal and lateral area, where the toilet is made for mechanical reasons

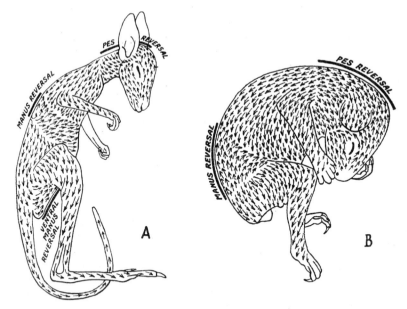

Fig. 41. Direction of hair in young animals from the mother's bag is already in agreement with the hair direction of the adults which agrees with the direction of toilet movements of the animals. A. in the Wallaby *Wallabia greyi*, B. in the Koala *Phascolarctus cinereus* (from WOOD-JONES).

towards the tail, does the hair lie tailwards in the embryo as well as in the adult (Fig. 41).

In both cases, therefore, the hair direction is determined before the fur has grown and before any actions of the already born animal could possibly determine the exceptional hair direction of these areas. WOOD-JONES[94] concludes:

"For undoubtedly there is a higher sphere of inheritance and it must be sought, not through a microscope, not even by the most nicely adjusted experiments, carried on for a short time in the short life of one human being, but by a survey, incomplete though it must necessarily be, of what Nature, with time unlimited at its disposal, has effected among living things. We may be running the risk of becoming blind to the whole range of the possibilities of inheritance, if we concentrate too exclusively on the minutiae of the means and modes by which, in certain cases, it appears to be effected."

We believe that dauermodifications as well as pre-adaptations, pheno-copies perhaps even pre-inductions (in WOLTERECK's[93] sense), govern the changes in form, function, and behaviour of the organism in varying environments. In no case is the reaction a simple physical one. The organism responds to the environment on its own. The fact that the organism and its way of response are the real point of interest has often been forgotten by those who measured the environmental factors as well as by those who counted only genes, chromosomes, and single characters.

## 8. Outlook

Every biologist admires the great achievements of modern genetics and respects the strict methods of genetical thought, the logic of its morphological and experimental basis. But the undisturbed peace of mind of the geneticist is not a final aim of biological research. The fact that little morphological or physiological basis for environmental hereditary-induction, for heredity of morphological, physiological, or behaviouristic reconstructions resulting as reactions of the organism to enforced environmental conditions is known, and that no bridge has as yet been discovered for the morphological understanding of the parallelism between phenotypical and hereditary geographical variation under the influence of the same factor, does not justify the rejection of such possibilities. That in some of these problems cytoplasmic influence counterbalancing the gene-chromosome mechanism may be made responsible has by now been realized at least by some outstanding geneticists[95]. As long as students in such important branches of biology as ecology, zoogeography, comparative anatomy, palaeontology, taxonomy, and experimental morphology, often feel forced to declare that, following their own methods, some kind of Lamarckian theory "is probably the only intelligible theory of a natural evolutionary process ever advanced[96]," it is the duty of the geneticist to reconsider the actual limitations of his present methods and to help his fellow scientists to understand these problems in which we are all equally interested.

At no time has cooperation seemed more necessary among all those scientists interested in understanding the living organism and the interactions of all intrinsic as well as extrinsic factors influencing it. To modern genetics we owe the abolition of that type of thinking with the aid of which understanding of the living organism was looked for by theories based entirely on deductions from insufficient facts alone.

Modern ecology is trying to gather reliable facts into a comprehensive synthesis. Cooperation is most urgently needed for the interpretation of these facts as well as of those gathered by genetics (in the broadest sense, including development mechanics, etc.). And we doubt whether any one side will be able to understand the living organism by means of its own methods on its own right alone*.

---

* See note g page 268.

## VII
## WHY HUMAN ECOLOGY?

Human Ecology deals specifically with a number of minor problems, besides throwing light on several grave and fateful issues which will decide the future of the human race.

ZINSSER[1] in a vivid history of the multi-faceted appearances of typhus, has laid stress on its ecology which was transformed by altering habits – especially by changing the shirt at least once each month – with the correlated change in the life-conditions of the lice transmitting that disease. Other epidemics are deeply influenced by weather, as BURNETT[2] has shown for a number of viruses. Changed coaptations are more important for changed relations between organisms, as is illustrated by ornithosis, including those of every intimate contact of man with infected birds, be it with pets like parrots (psittacosis) of with fulmar feather-picking by the fisherwomen of the Faroers. We have long known of the relation between weather and certain diseases. Some heart conditions, e.g. depend largely upon physical weather conditions when these differ too greatly from the optimal weather of the warm and humid tropics within which man evolved a million years ago.

In Israel, agricultural crops have depended upon the proper incidence of first (malkosh) and last (yore) rains ever since Biblical times. The seven fat and the seven lean years of agriculture (Genesis) gain in importance all over the world in the same measure that dynamic meteorology demonstrates the existence not only of long, centenary cycles of heat and cold, or of rainy and dry weather, but shows that within many regions more or less rhythmic shorter cycles of several years' duration of one weather follow several years of the other. In applied entomology WELLINGTON[3] has recently shown that two enemies of the fire-forests of Canada *(Choristoneura fumiferana* and *Malacasoma disstria)* increase and decrease in alternating cycles which correspond more or less to regular spring changes in the passage of warm- and cold-weather fronts over the country, the effect of which is heightened by the behaviour reactions of their caterpillars during a certain critical period. In a similar, often rather complicated way, many agricultural crops and their pests are influenced by weather. Changes in crop rotation, introduction of new varieties, penetration of new pests or fungi, new methods of soil preparation, etc. are influences of equal importance.

The types of growth of human settlements and their location, the preference of the western direction of the compass for the more wealthy residential suburbs, climatic comfort etc., are also often treated in Human Ecology.

The entire history of human civilization and technology is a history of human ecology in so far as it concerns changed relations towards nature, its domination and exploitation. Here, however, we will restrict ourselves to the few basic problems of human survival bound up with the concept of maintenance – maintenance of human populations and maintenance of the nature which surrounds human existence and upon which the latter's survival depends.

The first problem met with in these fields is the change of populations. The rate of increase of an animal (and such is man, despite many peculiarities) is correlated in nature usually with stable, unstable or directively transforming environmental conditions. Each population density requires certain conditions for its existence. When the population increases, the optimal niches for breeding, for hibernation, etc., are increasingly occupied and a steadily rising majority is exposed, especially during the sensitive period, or at the most unfavourable seasons of development, to greater harshnesses of the environment.

With man this has changed fundamentally. Despite all his progress in civilization and technique, man formed part of the natural fauna until the very recent past, in that his great natality was equilibrated by a corresponding mortality of infants and of mature but pre-senile adults. Today epidemics have lost their decimating strength. Conditions for infants, especially during the lethal summer days which cause mass toxicosis, have improved. Throughout our entire lifetime preventive medicine – with its sulfonamids, antibiotics, the astonishing advance of surgical technique, etc. – has lowered mortality enormously. The weak are hospitalized early and encouraged to carry on to reproduction and old age.

This implies a revolution not only in the age structure and in the social conditions of the population, but also in its genetical composition. The growing fraction of the population in need of social help from public funds has a most unfavourable influence on economic possibilities for positive eugenic measures, for public education, etc.

In genetics the situation is even more complicated. All those less viable recurrent mutations, the great majority of which were weeded out by natural selection in earlier days, are not only now brought to old age but also to reproduction, increasing the number of the mentally feeble, of those with a constitutional debility such as tuberculosis or serious diabetes, of the constitutionally weak and socially unfit to maintain themselves.

R. Cook[4] stresses especially the regressive tendency of the intelligence quotient (I.Q.) as a measure of the mental standard, which he calls an erosion of genes. The Report of a Royal Commission in Great Britain[5], and even J. B. S. Haldane – who tends to an extremely environmentalistic position in human sociology – agree that the I.Q. of England shows a clear trend of intelligence decline of one to two points per generation (l.c.). The entire argumentation should, of course, be read either in the Report or at least in R. Cook's book. Professor Burt concludes "that in a little over fifty years the number of pupils of "scholarship" ability would be approximately halved and the number of feeble-minded would almost be doubled." This view is almost general amongst competent observers and human demographers: that the intelligence level of future populations will be not only poorer in percentage of intelligent individuals but much richer in people with a low I.Q.

This means that a physically, mentally and socially weaker, but rapidly growing population will have to struggle for existence in an environment which is becoming increasingly unfriendly and unsuited for the production of sufficient food and work. The rising menace of automatization will deprive still more people of their bread. It is true that "we are part of a larger scheme of things than we can understand. We cannot describe it . . ., for we ourselves live in this whole and by it, and are wise only in the measure to which we make our peace with it[6]."

Nobody would propose turning the clock back and restituting the selective mortality of earlier times. But then there is only one possible solution, namely restricted natality, planned under competent genetical supervision, and this not by one or some nations, but by the entire world.

This brings us to the problem of optimal density. Optimum concepts are essentially connected with stable populations, and are favoured in Anglo-Saxon demography for the desirable density of a given population in a given territory.

This concept was first applied to fishery. The danger of overfishing the halibut in the Newfoundland banks or of the whales in the Antarctic seas, has given rise to the term "optimal catch"[7], the catch which implies the highest possible utilization of the fish- or whale-populations without endangering future catches, thus permitting full and permanent rejuvenation of the fished population.

For human populations A. Carr-Saunders[8] and others have long advocated, following Malthus[10], that the impending overpopulation of the earth can be avoided only by a policy of stabilizing at optimal density the population of every country and of the world as a whole. This optimal population-density is above a population-low, at which communications, food production, etc. are insufficient to assure full utilization of the country's natural wealth, considering the state of civilization, but below the level where utilization of the national wealth becomes uneconomic – when, for instance, the individual has at his disposal less soil than he is reasonably able to work. Beyond this point food production becomes insecure and industrial production unstable, because of their high dependency upon foreign markets. In our world, considerations of national security have also become weighty. Many problems of a "synthetic optimum of population" have been discussed by I. FERENZI with a view to practical implications[9].

Obviously, optimal catch and optimal population density have entirely different meanings from the point of view of population theory. In optimal catch we are dealing with a stable, oscillating population. In a growing fish-population the concept would be devoid of meaning, because this optimum would depend upon the actual position of the fish population with regard to the logistic curve – that is, it would change enormously, eventually, in short intervals. In more or less stable populations the optimum is an empiric average catch, which would combine relative maximal utilization with relative maximal protection of the future catches. If the empirical average is wisely chosen, it may represent an excellent compromise, assuming that the unfished population excess in years of greater abundance will make good the results of overfishing in years of scarcity.

In contrast to this static-population concept, the optimal density of human populations has a different and very dynamic meaning. It would be almost superfluous to apply a policy of optimal density to a stable human population. Such a policy can concern only growing populations. It is uncontrolled growth, often sudden, of before heavily reduced infant and pre-senile mature individuals, combined with the lessening of old-age mortality, that worries demographers. It is the rapid population growth of China, India, the Arab world, etc., where the introduction of modern preventive hygiene and, at an earlier date, colonial economy, have suddenly and enormously raised the percentage of surviving infants and of the old-age classes. This increased expectancy of life, even without any increase in the birth rate, effects a rapid population growth with no corresponding increase of food production or food supply. If there is an optimal population it is somewhere about the middle of the logistic curve of population growth, certainly not along its upper asymptote.

This impetuous tendency of population growth towards the upper asymptote of the logistic curve has many serious consequences for the world's future. Probably the most important one is the growing numerical preponderance of regions which in the past had little voice in the ruling of this world. At the same time the increase of economic misery is in precisely those regions where people have for long lived on the borderline of starvation. There is not the slightest doubt that from the point of view of a rational population policy conceived on a world basis, it would be imperative to repress further rapid population increase in those regions which are already unable to maintain their present population on a satisfactory health level, until economic conditions, and especially food production, have improved to such a degree that further individuals may be maintained at a comfortable health-minimum of existence.

So far humanity has shown itself woefully unfit for the rational mastery of its own future. There may even be an objection in principle to any project of population policy, namely, that the pressure of the need and of the misery will in due time enforce their own solution by creating improved methods of production. Even if that should be true, it would be achieved only at the cost of much misery during the transition periods. We have little confidence that humanity – which will reach four billion individuals about the year 2000 – will have the wisdom, the courage, and the decision in time to curb

(a) the unhealthy ever-growing increase of numbers,
(b) the steady and progressive qualitative degeneration of the same population.

Wisdom and sacrifice are demanded on all sides to overcome the dangerous, nay, fatal, laissez aller policy which dominates present population trends together with dogmatic wishful thinking of professionals, separated by their dreams from the real world.

Nobody with ecological training can underestimate the vital importance for all human problems of this steady growth of an ever qualitatively deteriorating population. We must come to the conclusion that this problem is certainly even more important than the menace of the H-bomb – if it is not solved by it.

R. C. Cook leads us, in his highly readable and commendable book "Human Fertility" through many of these present and looming miseries in many parts of the world. It is well worth our while to follow him in his deliberations briefly[11]:

"Unbalanced and unchecked fertility is ravaging in many lands like a hurricane or a tidal wave. In Puerto Rico, Egypt, India, Italy, and Japan, rampant fecundity has produced more hungry mouths than can be fed. The scramble for bare subsistence by hordes of hungry people is tearing the fertile earth from the hillsides, destroying forests, plunging millions of human beings into utter misery. Even where unchecked fertility has not impelled human beings to a scramble for survival, what is happening today has other sinister effects. In the United States, England, and every Western country, misplaced and badly distributed human fertility is leaching away the inborn qualities of tomorrow's children. This biological "erosion" is insidious in its action. No barren, gullied hillsides meet the eye, but in the foreseeable end such erosion of the biological quality of the people may, after no great lapse of time, result in disaster... number and kind of people – that is the crux of the crisis...

"Until a century ago the human struggle for survival was like that of any other species. Death was the major selective force. And it was a very active force, for only rarely did a majority of the babies born live to become parents; often, half the babies died before their first birthday. This pattern of reproduction was cruel and wasteful, but it was the basis for maintaining genetic fitness over the long pull. It was the universal pattern of human survival until death control began to be effective, about 1850. Then a great change occurred. Man, the toolmaker, whose gadgeteering had ranged from the spear, to the wheel, the decimal system, the microscope, the steam engine, to the dynamo, turned his restless mind in a new direction. He began to invent vital tools, tools concerned with the human life process. Vaccines, sanitation, aseptic surgical and obstetrical techniques, public health measures, fertility control, antibiotics – these are only a few of the new tools that make this age unique ... For the first time, a biological species began to control death; "Who is born?" "Who died?" "When?". The "when" in the equation has shifted. As a consequence, human numbers and kind are changing radically ... Few people today realize that the new powers over life and death are more portentous in terms of good and evil than any other single advance the human species has made."

Cook gives us as an illustration the island of Puerto Rico, which in 1898 became an American responsibility. On this island of 3,400 square miles the population had grown by 1928 to 1,500,000, by 1950 to 2,200,000 people, and this in spite of heavy emigration. The birth rate at that time was, at $50\%_0$, among the highest of the world. Modern health service took care that the mortality among those born was minimal. There is actually no way out other than voluntary birth limitation to check the further growth of the population on an island where every square foot of soil is utilized, and whose national economy is desperately negative, as Puerto Rico requires heavy importation of foodstuffs.

G. Winfield[12] says exactly the same of China, which he knows well. There every inch of soil is cultivated and conserved by an age-long tradition of agriculture. On the one hand there is unchecked fertility combined with modern health services, and on the other hand, there are severe recurrent famines, with no transportation available to bring food into the starving areas – and most of the people anyhow live normally on the borderline of starvation.

"It is obvious that the first objective of the medical health programme must not be the simple, natural one of saving lives; instead, it must be the development of means whereby the Chinese people will reduce their birth rate as rapidly as modern science can reduce the death rate ... It will seem rank heresey to propose that during the next twenty or thirty years not even severe epidemics in China should be attacked with every means available to modern medicine ... Existing misery and poverty can be permanently eliminated only when there are fewer, healthier people, with longer life expectancy and greater economic security. The future welfare of the Chinese people is more dependent on the prevention of births than on the prevention of deaths."

Thus is the message of this Christian physician and missionary.

Exactly the same conclusion is valid for India, for Egypt, and for many other countries. It is true that countries like Sweden or Canada can attack their (relatively) easy problems with easy solutions, which always include the spread of knowledge on birth control. We see in Japan an almost man-made problem rising before our eyes: – The American post-War occupation inaugurated modern health service with its fall of mortality, unaccompanied by a decline in natality. Emigration is limited. Consequently a rapidly rising population has to apply mass voluntary birth control now! In Italy the present trend of declining natality gives an opportunity, for the first time in eighty years, of an equilibrated population, if the present high rate of emigration is maintained. Without a lowered rate of natality even heavy emigration was no remedy for the population pressure.

Along with this hopeful development in Italy, we wish to point out the one real achievement in this direction in our times – the response of a noble, wise and courageous people faced with a full-sized population catastrophe. We here cite COOK[13] in extenso:

"In 1670, the population of Ireland was approximately in balance at about a million. Then the potato magically put into the mouths of the Irish people ample food to fill their stomachs. Numbers increased at an astonishing rate until subsistence again became marginal. By 1800, population had trebled. The census of 1821 gave 6,802,000, that of 1841 8,175,000, and the estimate of 1845 8,295,000 Irishmen. Rebellion, restrictive legislation, absentee-landownership, starvation, and demoralized social conditions induced by pressure of population on the land and by limited industrial opportunities, caused emigration to increase between the years 1815 and 1845. Many went to England's expanding industrial areas. The few who could afford passage came to America.

"Misery, starvation, and disease hit with cyclonic fury during the potato famine years, 1847 to 1852. That incompetent administration and speculation in grain added to the horror of the starving time is a shocking postscript to a ghastly situation. Death and migration reduced the population by nearly two million between 1846 and 1851. Long before the famine years, a drop in the birth rate had been recorded. But the brakes did not go on fast enough to avert disaster. From 1821 to 1831, the rate of increase of the population was about 14 per cent; in the next decade, however, the rate of increase fell to 5 per cent. The drop in the birth rate seems to have been caused partly by those same repressive factors which encouraged emigration, but mostly by a shift in customs of marriage and land tenure. The change in marriage customs which favoured late marriage and celibacy of both sexes seems to have dated from the early part of the nineteenth century. Today late marriages – and few marriages – keep the Irish birth rate among the lowest in Europe.

"At about the same time a movement favouring consolidation of farm holdings through marriage, so that the land would not be divided into unworkable bits, was adopted by the country people. This new accent on land tenure as the basis of marriage began to affect the Irish birth rate about 1820. So did the dowry, the kinshid system, and many other social customs."

It is of the greatest interest that all these changes of customs which brought about the lower birth rate arose spontaneously from the peasantry. The political leaders of the Irish movement and the clergy were always strictly against any birth reduction. By this revolution of the common people, without any organized leaders, the population of over eight millions is today down to 4,385,000 (slightly over 3,000,000 in Eire). Emigration alone could not have brought about this change without the corresponding birth reduction. There are still poor in Ireland, but the country as a whole is as healthy, as progressive in education as could be wished. The country has not only won its political liberty, but is the one nation in the world which has found a solution for the menace of the population problem.

The ecological observer will, however, stress the fact that birth reduction tendencies began some decades before the famine years. This means that the people had mentally and morally already found its way, almost instinctively – and this was and still is in our time a difficult way personally, with many privations and restrictions, but with liberty for the people – before the catastrophic famine years enforced the population reduction for a short while at least.

The courageous Irish people have learned their lesson, and they have – in contrast to human habit – not forgotten it once the years of the trouble have passed. This lesson can be learned by other peoples as well, that public information work with regard to birth control can be begun within the menaced areas before it is too late. Success will not always be evident immediately, but small beginnings may later on help towards rapid development, when catastrophic events will cause the many to travel the road opened up by a few in earlier days.

We shall now turn our attention to the other side of the same problem, to the steady increase in the unfriendliness of the human environment, with its unhappy implications. We shall commence with a discussion of the age-old importance of environmentalism for human civilization.

In the past century, geographical environment has often been stressed to the extreme as the dominating factor enforcing and forming social, political and cultural life[14]. RATZEL[15] and HUNTINGTON[16] are outstanding fighters for this view, Miss SEMPLE[17] following more moderately. There is, of course, no doubt that geographical conditions as a whole largely determine social organisation and social processes, and this in proportion to the primitiveness of the human organisation. The isolated families of tribes inhabiting primeval forests are conditioned to their means of existence, and primitive agriculture, brought to a higher level by the great irrigation civilisations of the Indus, Euphrates and Tigris, as that of the Nile, tends to social aggregation. Once man has reached a higher social level, his dependence upon topography becomes less conspicuous, yet persists to some degree. Many are the streams offering similar facilities for organised irrigated agriculture which has not been similarly exploited by the inhabitants of the area. The primitive Delaware Indians used the wood of the forests only for cooking and primitive heating. The modern Pennsylvanian has a choice of wood, coal, oil or electricity for heating, and in addition almost all of his activities are bound up with the use of one or more of these.

The same geographic environment holds a very different meaning for peoples of different character, especially where industrial utilization is concerned. The mountains of Switzerland and of Andorra, and the wide grass plains of Russia exemplify this point. The unfavourable ports of Tyre and Sidon and their most unfavourable hinterland certainly did not make the Phoenicians into great seafarers and colonisers – their innate energy and their commercial ingenuity were the determining factors. None of HUNTINGTON's contentions of the exclusive culture-promoting values of temperature or humidity has withstood reasonable analysis. People are not pessimistic because they inhabit a small and limited territory, and dwellers on the great plains are not happy because they may roam over wide areas, as RATZEL maintains. Are the Danes really more pessimistic, and the Russians happier than other people? Her geographical analysis has not prevented Miss SEMPLE from including the Saharo-Sindian deserts and the Irano-Turanian steppes into the Mediterranean. STRYGOWSKI & BODENHEIMER have shown that no cultural analysis of the Eastern Mediterranean hinterland is possible without taking these bio-geographical aspects into account.

We must beware of blind admiration for sweeping, onesided theories, like those promoted by HUNTINGTON or MACCOWAN, about the geographic determination of the origin of the three western monotheistic religions. The soft, rounded hills of Samaria and Galilee, the rugged hills of Judaea, and the wide spaces of the Arabian desert did not determine Judaism, Christianity and Islam. As if there were not plenty of homologous topographies all over the world which have never produced similar, or even any, religions!

The location of London, Paris and Rome – the heart of historical and modern empires – were not chosen under widely different and primitive conditions because they were favourable sites for these empires. Geographical causes determined the choice of their sites as suitable for the establishment of small towns, under very limited conditions. Coastal peoples are not forced by their environments to become important seafaring nations, just as nobody in history is ever forced to become important. But if such people take to sea-fishing, and if they possess an innate energy, they may develop it in the direction of the sea.

With due cognizance of the great importance of the geographical environment even for the highest social structures, we must deny it the dominating role as the truly deterministic force in the shaping of human civilization. If this is true of geographical factors in their entirety, it must be stated still more emphatically for any isolated geographical factor, such as temperature. We are far from any adequate analysis of the impact of geography upon the progress of civilization, culture, and science.

Totally different is the purely geographical approach of G. TAYLOR[18] who has followed in detail the progress of white settlement in Australia and in Canada. He comes to a clear environmentalistic conclusion: – a given soil and climate will prescribe the relative lines of settlement which are not possible on any other lines of equal population density, the so-called econographs, which indicate the lines of normal empirical population density possible under the given environmental conditions of today. Any change of these conditions will proceed in the future also along these lines of equal density, except where new projects or wealth is evolved or discovered, such as irrigation dams in areas where they are uneconomic today or great deposits of uranium ores. But the new calculation of the econographs will automatically give expression to the changed conditions.

We now proceed to man-made deteriorations collectively known as erosion. Erosion normally is part of the natural processes forming fertile soil from rocks and spreading it by wind, water and frost over larger surfaces. Such forces building up the soil and destroying it both are at work even in the absence of man. In his absence natural processes tend to become stabilized in a way usually favourable to the maintenance of fair vegetation. These processes proceed usually in the Lyellian tempo of geological processes, meaning that they are slow in both directions. Even historical man had already forced changes in the tempo of these processes towards more rapid soil destruction than could possibly be balanced by the slow building-up of good soil.

It is the tragedy of modern man that the rapid development of agricultural science has induced an avalanche-like increase of soil destruction by intensive, unwise soil exploitation, because no consideration was given to the permanent protection of soil and vegetation.

Modern man certainly remains a part of nature. Thus it is incorrect to say that he destroys the natural equilibrium. Destruction by man is part of the natural processes of the present, and the destruction of soil and vegetation in man's steps is also "equilibrium" in the sense of our earlier definition. But it is no longer a condition of maintenance – it is a condition of deterioration leaving less and less soil of good natural built-up with natural fertile vegetation, with less possibilities for future agriculture in the years to come. The destruction of soil and vegetation – both of which are the natural setting for fauna and man – grows in dimensions like the proverbial snowball once the processes of erosion gain in time and space[19, 20].

Man's drive to utilize the day and to exploit the possible benefits of the moment without adequate heed to the future productivity of the soil may ultimately lead to his destruction.

Last autumn we saw this contrast vividly. A trip through Tuscany gladdened our eyes with the sight of a landscape intensively cultivated by farmers for thousands of years, ever inch of cultivable soil still green and productive. It has produced rich crops for many centuries and will continue to yield for uncounted centuries to come, as long as proper care is given to the conservation of the soil, to proper crop rotation, etc. as before. A few days later we landed at the aerodrome of Brazzaville, in an area which a few years before had been part of the incredibly lush tropical rainforest of the Congo – today it presents a picture of a desolate desert. As far as the horizon reaches, not one herb, not one blade of grass grows in this man-made desert. HEIM[21] tells a similar story from the area of the earlier primeval forests of Madagascar, PHILIPS[22] of areas in East Africa – all man-made deserts in the originally fertile areas of the tropics.

We know of Oklahoma, in the famous dust-bowl of North America, where unwise crop rotation led to having large stretches of fertile soil blown away, leaving behind a man-made desert. The Americans, however, were the first to visualize the damage caused by erosion, the first to become erosion-minded. Much serious research, many minor experimental and extension plots were begun there. We therefore asked a young scientist from Oklahoma, when we met him last year, about the present state of soil-erosion in his home-state. He answered, with a shrug of his shoulders: "Who cares today in Oklahoma about agriculture? Oklahoma today is an oil country, and nobody bothers about anything else!" A few years ago, the members of a scientific desert symposium visited some of the southern states of the U.S.A. They all wondered about the rapid progress of man-made deserts in these regions. The simple protection of fencing was in certain areas sufficient to produce rich vegetation surrounded by overgrazed, eroded land.

This is the great problem of the Australian continent, where overgrazing in the fat years created a "desert" for the lean years, not to be restocked adequately even during the next series of fat years.

We could continue with a long, long list of similar phenomena almost all over the world, especially in arid and semiarid regions, but what has been mentioned here is sufficient to demonstrate that man can create and has created man-made deserts from wide areas of once fertile land which now produce nothing and which demand entirely uneconomic means in order to become fertile again.

Prehistoric and primitive man were both part of the natural landscape. Man satisfied his needs by collecting berries, fruits, roots, and hunting insects, shellfish, fish, reptiles, birds and mammals, in modest measure, without ever abating the abundance of those plants or animals which served as food. Men destroyed only a tiny fraction of the fruits and seeds available, often not even destroying the seeds. Man was not more dangerous to the continued existence of his animal food than any other big carnivorous species. The small number of palaeolithic men had no influence upon the fluctuations of his food and prey-populations, which were subject to heavy fluctuations by other environmental factors. This is easily proved by analysis of the alternating fluctuations of Fallow Deer and Gazelle in the palaeolithic caves of Mt. Carmel[23], which were not fluctuations of taste, but of game available around the caves. Man had not more force to change the landscape which he inhabited than had any one of his fellow large mammalians.

Even the early development of agriculture and later of animal husbandry were unimportant events in this respect. The few scattered patches in the natural steppe where the first sown areas were established, were only local events of minor extent. Greater were the consequences of the first organization of intensive, irrigated agriculture in the great river valley of Mesopotamia, India and Egypt. These were local improvements of moderately fertile loess by irrigation or utilization of regular annual inundations which spread water and fertile soil over densely cultivated, but restricted areas. We now realise that this prolonged intensive agriculture was only possible because of continuous alternation of the fields grown infertile and their exchange for fallow land which had lain uncultivated for many years. The incredible wealth of agricultural Mesopotamia in ancient times is a gross exaggeration. Never did that apparently dense net of irrigation channels co-exist. One channel was opened when the earlier one was silted and when the land to which it led had grown sterile. New channels were then opened to new fields for a few years, and again new channels were opened.

Agriculture remained, until the impact of modern science, a purely empirical profession with very slow progress. The first and most important changes – from the primitive tools of agricultural work, combinations of wood or bone with stone, to bronze and iron tools – was very, very slow. Thus also was the growth of driving force available for agricultural work: wheels for water-carrying and for chariots, mills; draught animals were, in spite of their importance, means so old that no signs of their invention have been preserved.
Forestry. Deforestation followed by goat-grazing and charcoal production from the roots of the felled trees, thus preventing reforestation and regeneration, led to complete and rapid erosion on hills and mountains. This is the history of large-scale destruction of once fertile land in all Mediterranean regions from Spain to Italy and Macedonia and Palestine since early history. Bare rocks remained and a few patchy soil relicts on which even the goat could scarcely subsist. It is wrong to burden the goat with the main culpability of this process. The most injurious mammal was and is man, who cut the wood for his use as timber, who made an export of charcoal from arid countries, etc. We saw the first steps of such rapid deforestation in 1942 in Iraqui Kurdistan: Along the newly built highways entire slopes of the wonderful Kurdish oak forests were cut down and stapled for transport. Fortunately the local government understood the situation and took energetic preventive measures to stop this crime. We can only hope that this zeal is still maintained. If not, we will see the usual picture of history: – every rain will wash down tons and tons of fertile soil, until only the bare rocks remain a few years later.

One cannot assume that a devastated compact country-side can be merely replanted and regenerated. It is not that the goats effectively prevent the growth of seedlings, so that none has a chance to grow to maturity. Before the goat begins its work, the lack of shade, the poor deteriorated soil have killed the youngest seedlings. It is now about a hundred years since the idea of the "dauerwald" (continuous forest) of MÜLLER was widely accepted. Mixed forests know no general catastrophes by insects and other pests. All ages of trees must co-exist in order to prevent a tree-less stage in forestry. Then, and only then, all or part of the fully grown trees may be cut and the forest, nevertheless, will continue to exist as if nothing had happened. The writer has recently had the opportunity to see a well-preserved dauerwald in a prevalently monospecific forest in Finland. Timber export is the greatest treasure of that wonderful country. The pines there grow very slowly and to no considerable size because of the cold soil. But despite this slow growth and heavy export, Finland has been able to increase its wood reserve considerably in the last years. The climate is unfavourable but the wisdom and the care of the people brought about this happy condition. The country could easily have sold all its forests within a few decades for newsprint to all the world. Yet the writer has never seen a nation so ecologically-minded in its entirety, so considerate of the welfare of its sons and grandsons.

Fully eroded hill regions in semi-arid climates offer very difficult problems. There, complete erosion often dates back to 3000 years ago. This does not imply that the evil has been stamped out. It is interesting to have the expert American judgment on the timber aspects of such a country (a small one): From the point of view of economics it is ridiculous to replant the bare hills with poor and crippled pines which will not yield in any near future timber of any value at all. But if the limited irrigated land of the plains were to be planted with selected, suitable trees, they would grow rapidly and well, producing excellent and valuable timber. This is, of course, entirely correct from the viewpoint of the timber expert, but it is very wrong from the points of view of a small nation which intends to exist and to maintain its children now and in the future. No wealth of timber production will compensate for healthy agriculture on this irrigated land, which is the backbone of the agricultural produce of the country.

To return to the crippled pines on the bare hills. True, this hill afforestation is almost worthless in its first generation. But in time it helps collect, build up and retain much soil. By the yardstick of generations, this means that in the future valuable climax forests or fruit orchards can be planted and grow there. In addition, much rain water will be retained in these soils, which may change the agricultural aspects of an entire country considerably. This point of view is precisely the ecological one to which the utilitarian greed of our days is strictly opposed.

Even in the humid tropics deforestation without rejuvenation may often occur. J. HEIM (1951) has told the tragic story of Madagascar where, in a few decades, the forest area had been reduced from twenty to three million units. J. PHILLIPS (in "Biology of Deserts", London, 1954) reveals a very urgent problem concerning the sub-arid margins of tropical forests in East and South Africa. When tsetse-fly areas are made tsetse-free, overstocking by poor livestock (because of the need to purchase wives) very often occurs, bringing about the curse of rapid erosion and dessication, with the progressive spread of such new foci. PHILLIPS pleads to make only such areas tsetse-free where the proper use of the soil and its conservation on an organized basis are assured. Overstocking with sheep has transformed wide areas of scrub-steppe in Australia into eroded deserts which are lost for pasture. It is important to realize that not only semi-arid regions are especially sensitive to erosion and desert-formation, but that to some degree this may occur in almost any climate, whenever the management aims primarily to make profits to-day, regardless of the destruction of nature or of the maintenance of the natural equilibrium, including agriculture, forestry and fishery, as the primary basis of any planning.

Sea-fisheries[26]. In sea-fishery man's activity likewise had no importance whatsoever upon the fish populations until the improvements of modern ships and marine traffic here too transformed a craft into an industry. But here early ecological methods were applied and prevented many disasters.

The fisheries of Northern Europe have their natural fluctuations, of which we know little except that they exist, and are in an oscillating equilibrium of saturation. Some decades ago, as HJORT (1934) states, an enormous increase of potential catches was induced by the inauguration of deep-sea fishery methods. A slow beginning was followed by many years of rapid increase, which neared – towards the end of the typical logistic curve of population, etc. growth – another oscillating equilibrium at a level, however, of much higher catches. Restrictions became necessary in order to prevent overfishing. This was a new concept for marine fisheries, but in various places rich areas of fishery had shown progressive depletion which made further fishing there uneconomic. The ecological answer was the new concept of the Optimal Catch.

The experience of the First World War interruption of fishery in the German sea had shown that intensive fishing does not necessarily lead to the destruction of the fish populations. It was observed, however, that at the end of this period of interruption the higher age classes of fishes were much smaller than usual. Higher population density had thus led to a considerable reduction of the rate of growth. This is an indication that the seas possess oscillating equilibrium of fish-weight, where a greater number may be replaced by smaller mass or weight, and vice versa. Every attempt to improve fishery conditions must, therefore, be based primarily on studies of the prevention of unnecessary losses in these almost stable fish populations under the pressure of regular intensive fishing. This includes principally the protection of eggs and of the young brood during the spawning season, avoidance of the loss of high percentages of young and middle-sized fish still too small to be sold on the market, which are thrown back into the sea, largely after they have died. The loss of the spawn is of course easily avoided by prohibiting the fishing of adult concentrations during that season. The second type of loss is also easily avoided by regulating the mesh-size of nets. If these are too small, too many of the young fish which have not reached fishable age are destroyed; if these are too big, the catch is smaller that it could and should economically be. The determination of the proper mesh size for each specific Optimal Catch has by now become a routine matter. These few hints concerning an ecological problem and its key-solution may suffice to explain the ecological approach.

In parts of the Mediterranean Sea the relatively low density of fishes is well-known, even if its reasons are not yet adequately understood. It may even be that local, seasonal aggregations of certain fishes may occur. It is clear that under present circumstances no expert could assume that the doubling of any regional fishery fleet would produce a double catch. So long as the present fish populations have not been studied intensively and extensively by fishery-biologists, the aspect of such an undertaking is quite different. If – and this seems to be the case – the present fishery works under conditions of oscillating equilibrium of relative local saturation, any such increase of fishing intensity may easily bring about decimation and destruction of the regional stock, i.e. induce overfishing.

The ecological solution of the whale and seal-fisheries of the Arctic and Antarctic has also – unfortunately, very late – resulted in the international acceptance of an Optimal Catch. This time the protection of the breeding season was easily accepted, and no further protection of the young age-classes was required. The quantitative share of every one of the whaling nations is determined annually by an international commission of experts. A similar procedure has been established with the prawn fisheries on the Norwegian coasts.

The Newfoundland banks represent a classical object of Optimal Catch research. There a flourishing industry had its very existence menaced by overfishing. Ecology has proven in all these cases to be not a romantic return to nature but the only sensible way of maintaining the resources of fishery in the same numbers which permit the maintenance of an oscillating population in saturation and the establishment of fisher-intensity limits within which no danger exists for the maintenance of fish productivity.

Plant and Animal Husbandry. As we have stated before, agriculture was for long a purely empiric occupation which scarcely deserved the name of science. This changed rapidly around 1840, when the great progress in animal and plant sciences reached such a standard that their application upon practical agriculture was astonishing fruitful.

Since PRIESTLEY in 1774 discovered that the air in which a candle has burnt out is restored by vegetation, and INGLE-HOUSZ in 1777 pointed out that this process of purifying bad air occurs when the vegetation is exposed to sunlight, and SENEBIER in 1782 proclaimed the leaves as the essential organs of the plant, the intensive study of plant nutrition began. DE SAUSSURE in 1804 realized that the nitrogen within the plant is not derived directly from the free nitrogen of the air but from the soil, especially from the "humus". DE CANDOLLE in 1832 described the chemical mechanism of starch formation in the leaves, etc.

Then came the great discoveries of LIEBIG, DUMAS and BOUSSIGNAULT. LIEBIG (1840) exploded the humus theory, that of the origin of carbon from organic wastes. Actually, humus is not absorbed by the roots, the only source of carbon available to the plant being the $CO_2$ of the air. Carbon dioxide, ammonia and water contain within them the elements required for the production of all "organic" substances of plant and animal life. They are in the same way the ultimate products of the chemical processes of organic decay. BOUSSIGNAULT (1861) demonstrated that plants are unable to make any use of the free nitrogen of the air but derived all their nitrogen supplies from nitrates present in the soil. At the same time the minerals constantly present in plant ashes were recognized as essential for its metabolism and welfare. BERTHELOT (1876) and WINOGRADSKI (1890) showed definitely that the free nitrogen of the air is fixed in the soil by the activity of bacteria in the root nodules of leguminous plants. These are a few of the glorious discoveries in plant physiology and biochemistry which after 1840 rapidly transformed the face of empirical agriculture into a profession under scientific guidance, into a practice with science.

Before the impact of science certain achievements of no mean importance were already gained. We had known that wheat sown after wheat exhausts the soil rapidly and that it takes some time until the impoverished soil is sufficiently regenerated to yield normal crops. In parts of Eastern Anatolia only a few years ago a piece of land would lie fallow for up to nine years after having yielded one crop. In Iraq, on the contrary, one field might yield good crops for three or four consecutive years with irrigation, then remaining fallow for two or three years. But in this fertile region the streams year after year bring the most fertile soil with them in the Spring. The abundance of land in both these regions makes this system of long fallow intervals possible.

In Europe, the first great change in agricultural empirism was the progression from a steady biannual crop-rotation with one field sown, the other fallow, into a three-field crop-rotation, made possible by having two or three fields under different crops, the third only remaining fallow. Later, the scientific methods of preparing and preserving organic manure, the introduction of guano and nitrate of soda from the Americas, of potash, phosphates, etc. enabled the farmer to restore his depleted fields immediately, thus doing away altogether with the system of fallow fields. Mechanical improvements included drainage, soil conservation, contemporaneous drilling of seed and manure in rows, etc. Eventually the farmer learned that manure was still the basis of good farming but that it can and should be improved by chemical manures in order to produce heavier crops. The discovery of the nitrogen fixation by the roots of the leguminous plants led in mild climates to crop-rotation with two crops the year, of which one is usually a leguminous one such as fodder, especially where irrigation is available, and in eliminating entirely or reducing to one fourth in one or two years the fallow soil. Cutting, reaping and flailing or threshing were replaced by the huge threshing machines, the horse-plow by tractors, and still greater machines appeared in the combines. Every part of the work was highly mechanized.

The primitive condition of plant control thirty years ago is typical of every branch of agriculture – everything a generation old is definitely "prehistoric"! In Europe, the first impact of this high mechanization of agriculture came in good time to prevent serious loss in manpower in the rural areas, as the young people were leaving the soil for easier and more remunerative work in industry in the cities, or for better conditions in agriculture abroad.

In less industrialized countries mechanization produced a progressively difficult situation. Y. Volcani, the father of Palestinian agricultural science, said justly, that the mechanization of a farm is good and beneficial when and where the farmers have productive use for the time won in this way. When they save 90 days of work per head by full mechanization and have no productive work on hand for 50, 60 or 70 of the 90 days saved, it means the mechanization is not an economy but a monstrous luxury.

This leads us to the greatest and most unpleasant of all the changes which have taken place in agriculture – the farmer's mentality. In older times the farm and the soil belonged to generations, it being fully understood that the future offspring of the family would live and earn their bread there. The farmer felt himself one with his soil. This attitude is no longer present in most countries – in many areas of new colonization this relation never existed. Now, usually, agriculture is nothing more than a crop-producing, money-making industry. The industrialist calculates without thought or care for the soil's fertility a few years hence. Even if he realizes the consequences of his rape of the soil, he is ready to sacrifice it for his own benefit this and eventually next year. As long as additional soil in quantity was at his disposal, he did not even consider an alternative. Now that soil has become scarce or fully settled in most regions, he as well as the State are forced to give mind to the future. The merely mechanistic utilization, exploitation and rape which has destroyed enormous areas of once fertile soil cannot any longer be tolerated by the hungry, ever-increasing population of our planet. The early settlers of Oklahome did not know what they had induced. To-day no agricultural planning can be tolerated except that based thoroughly upon ecology. Even temporary and emergency planning can no longer allow for the destruction of natural maintenance to which soil and agriculture belong, as well as vegetation, fauna and man – now and in the near and far future.

A few examples will serve to illustrate the meaning of ecological planning. In a fertile region with irrigation a heavy sheep may be bred, which would be unable to find its maintenance on lean soil. Eventually the two mouths and eight legs of two lean sheep may find their food there easily. It is thus clear that improvement in a race may be valuable locally, but when brought to the wrong place, the benefit turns into self-destruction. We should bear in mind that generally, every increase in crop productivity is limited locally. Limiting factors of any region cannot be overlooked, even if they only appear at long intervals. Thus, in the very cold winter of 1930 almost all the appletrees in East Prussia were killed by severe frost, merely because the farmers had forgotten that such frosts occur regularly, even if usually at long intervals, in that region. Everyone knows of similar cases in his own country.

Having thus touched upon animal husbandry, a few other aspects of ecological thinking may be cited. The Italian bee was introduced into a country with its own well-adapted native bee-race. The bee, excellently adapted for Italy and for wide areas of the U.S.A. where no honey-bee had been before, is decidedly unsuited to a really arid climate[22]. It cannot possibly produce more honey there than the local bee, but the whole beekeeping industry of that country is so backward that nothing of this kind could be proven. The only reason given for this introduction was that the local breed was wont to sting. But at least two convincing experiments by private beekeepers there demonstrated that it is easy, within a very few years, to select non-stinging quiet strains of the native bee. There was thus no real reason for replacing the native bee by the Italian strain, which is probably ill-adapted to the climatic cycle, the diseases and the prolonged drought of the local climate. The future of a profession, or even of an industry, should not be endangered by such haphazard methods.

That same country once served as the home of millions of domestic fowl who produced a modest crop of eggs. Now this native fowl, bred upon the natural waste of the farm, has been replaced by a highbred Leghorn which is kept in industrial conditions of breeding from the incubator to the market, on expensive food which is either taken from the inadequate local crop and rebought with foreign currency (which is not available), or has to be introduced into the country with foreign currency. Egg production there has become a machine, the parts of which are brought in and which is only put together in this country – a utilitarian luxury in agriculture for which there is no excuse. The ecological method would be to study how the earlier, primitive way of hen-keeping and egg-production could be improved without creating an industry strange to the country and its economy. Eggs of the proper colour, taste and shell-thickness can be produced, although this requires more and perhaps more difficult work than the mere copying of the American fowl industry model in a country unfit for its introduction.

Milk cattle in that same region is bred from imported Frisian bulls. In this way the hybrids with the native cows have gained a fair rise in milk-production and retained a fair degree of immunity against many of the local diseases, but their life habits are quite unnatural. The fodder is often grown in other parts of that country, and the cattle never leaves the cow-shed yard. It is only recently that experiments have been commenced to keep cattle, grown for beef, on the pasture throughout the day. The first results are encouraging, provided that the problem of pasture production and preservation in that semi-arid climate is solved. This is a problem for intensive ecological study, but the effort is towards a "natural" solution, not comparable to that of if and how coffee may be grown in Mediterranean countries. It would undoubtedly be possible to grow and maintain that tropical tree under hothouse conditions, but such problems are uneconomic whatever their outcome. Thus, in the long run, the ecological point of view is also the most economical one. Whoever wishes to preserve for coming generations what we have received from our fathers, improving it to the best of our knowledge, has to abandon utilitarianism and mechanization as guiding principles – not necessarily as auxiliary ones – and turn to a sound ecological analysis of past, present and future problems in agriculture, correcting the evils so far initiated.

We are at a turning-point in food-production. There are no major problems of improvements, no problems of further intensification, no further problems of utilitarian mechanization. There is the choice between two basic attitudes, two philosophies of life. Any further progress of utilitarian industrialization will close the door to the possibility of turning back the wheel leading to destruction. This is our last chance to replace it energetically by ecological, long-term planning, carried out against all group resistance of vested interests. We need scientific analysis no less than the alternative philosophy, but in other directions. We could not better conclude this chapter than with MICHEL REMY's words:

"La terre épuisée refuse de continuer à nous nourrir. Il est encore possible à l'homme de choisir, avant que vienne le moment ou rien ne pourra plus faire que la production de la terre, répartie entre tous les inhabitants, leurs assure le strict minimum vital. Dans la lutte générale pour survivre qui commencera alors, une partie de l'humanité sera irrémédiablement condamnée à disparaître, au milieu du plus grand carnage qui se soit jamais vu. L'homme controle encore généralement le feu qu'il a allumé, quoi qu'il prenne parfois accidentellement des proportions gigantesques. Il peut encore arrêter l'incendie, et laisser la nature se reconstruire. Mais s'il ne le fait pas rapidement, il disparaîtra bientôt dans un fin apocalyptique. Penseurs et politiciens sont comme le timonier ivre au gouvernail d'un navire, qui, se croyant invulnérable, ferme les yeux devant les obstacles, ou il va finir se briser. Dirigé contre la nature, la science fera faillite."

The demographic problem of human fertility in a world which grows more and more unfavourable for human survival is largely one of human behaviour, of a better understanding of our proper reactions. The change of man's surrounding world by human activity is also largely conditioned by exogenous factors which can be changed in the future only by different human behaviour. It is man's tragedy that the splendid rise of agricultural science in the first half of the nineteenth century was so soon eclipsed by the deterioration of his surrounding world. There was a growth of agricultural knowledge without, as in many other fields, an adequate growth of human wisdom and horizon. The tendency today is to better the conditions of agricultural production without considering in how far today's benefit will impair tomorrow's productive capacity – to live well without asking whether our sons and grandsons will be able to eke their minimum existence from that soil. That is why ecology is needed for a proper understanding of man's greatest problems. And that is why no other science is at this moment more important for mankind's future than that of human ecology and the solution of its pressing problems.

# REFERENCES

## Foreword

1. LACK, D.,The Natural Regulation of Animal Numbers. Oxford. 1954, p. 275.
2. AYER, A. J., The Problem of Knowledge. London. 1956.
3. BODENHEIMER, F. S., 1957: Knowledge and Biology, in: *Studies in Biology and its History*. **I**. *63*.

## Chapter I

1. KORSCHELT, E., Lebensdauer, Altern und Tod. 2nd. ed. Jena. 1922.
2. FLOWER, S. S., 1925: *Proc. zool. Soc. Lond.* **247** (fishes); *ibidem*, *269* (Batrachians); *ibidem*, *911* (reptiles); 1926: *ibidem 1365*, (birds); 1931: *ibidem*, *145* (mammals).
3. PEARL, R., 1936: *Mém. Musée Roy. Hist. nat. Belg.* **2, 3**, *169*.
4. RUBNER, M., Das Problem der Lebensdauer u.s. Beziehungen zu Wachstum u. Ernaehrung. Munich. 1908.
5. PUETTER, A., Vergleichende Physiologie. Jena, 1911; also: 1921: *Z. allg. Physiol.* **19**, *9*.
6. PEARL, R., 1928: *Quart. Rev. Biol.* **3**, *393*.
7. ALPATOV, W. W. & PEARL, R., 1929: *Amer. Nat.* **43**, *37*.
8. MCARTHUR, J. W. & BAILLIA, W. H. T., 1926: *Science*, **64**, *229*.
9. RUBNER, M., 1908: *Arch. Hyg., Berl.* **48**, *260*.
10. COLLE, J., 1929: *Arch. int. Physiol.* **31**, *1*.
11. TERROINE, E. F. & WURMSER, R., 1921: *C. R. Acad. Sci., Paris*, **178**, *482*.
12. BARTHÉLEMY, H. & BONNET, R., 1924: *C. R. Acad. Sci., Paris*. **178**, *2005*.
13. KROGH, A., 1914: *Z. allg. Physiol.* **16**, *163*; *Int. Zbl. phys.-chem. Biol.* **1**, *491*.
14. PARKER, J. R., 1930: *Bull. Montana agric. Exp. Sta.* 223.
15. FILINGER, G. A., 1931: *J. econ. Ent.* **24**, *52*.
16. JANISCH, E., 1924: *Arb. biol. Reichsanst., Berlin*, **13**, *174*.
17. PHILLIPS, E. F., Beekeep. Annu., New York. 1928.
18. LUNDIE, A. E., 1925: *Dep. Bull. U.S. Dep. Agric.* No. 1328.
19. BODENHEIMER, F. S., & BEN-NERYA, A., 1937: *Ann. appl. Biol.* **24**, *385*.
20. CRAIG-BENNETT, A., 1931: *Philos. Trans. B.* **219**, *197*.
21. SMITH, L. M., 1937: *J. agric. Res.* **54**, *345*.
22. BODENHEIMER, F. S., Citrus Entomology in the Middle East. Den Haag. 1951.
23. MUNRO FOX, H., 1936: *Proc. zool. Soc. Lond.* *945*; 1937: *ibidem*, *275*; 1939: *ibidem*, *106*, *141*.
24. OGLE, C. & MILLS, C. A., 1933: *Amer. J. Physiol.* **103**, *606*.
25. GALINEO, M. S., *Ann. Physiol. Physicochem. Biol.* **10**, *1083*.
26. TITSCHAK, E., 1925: *Z. wiss. Zool.* **124**.
27. KOZHANTSCHIKOW, I. W., 1934: *Z. angew. Ent.* **20**, *489*.
28. GONZALEZ, B. M., 1923: *Amer. Nat.* **57**, *285*.
29. PEARL, R., The Rate Of Living. London. 1928.
30. Calculated from the data presented in: BALLARD, E., MISTIKAWI, A. M. & ZOHEIRY, M. S., 1932: *Bull. Minist. Agric. Egypt.* **110**.
31. WIESNER, B. P. & SHEARD, N. M., 1934: *Proc. Roy. Soc. Edin.* **55**, *1*.
32. DEEVEY, E. S., 1947: *Qart. Rev. Biol.* **22**, *283*.
33. MURIE, A., 1944: *Bull. Fauna Nat. Parks*. **5**. *481*.
33a. HATTON, M., 1938: *Ann. Inst. Océanogr.* **17**, *241*.
34. PEARL, R., The Biology of Death. Philadelphia and London. 1922.
35. DUBLIN, L. I. & LOTKA, A. J., Length of Life. New York. 1936.
36. BODENHEIMER, F. S., 1934: *Genus*, **1**, *43*.
37. BURGDÖRFER, F., Volk ohne Jugend. Berlin-Grunewald. 1932.
38. DOUDOROFF, M., 1936: *J. exp. Zool.* **72**, *369*.
39. BODENHEIMER, F. S. & DVORETZKI, A., 1952: *Bull. Res. Counc. Israel*, **1**, *62*.
40. KALABUCHOV, N., 1935: *Zool. J.* (Russia), **14**, *209*.

41. BODENHEIMER, F. S., 1937: *Quart. Rev. Biol.* **12**, *406.*
42. BODENHEIMER, F. S., Studies on the Honey Bee and Beekeeping in Turkey. Istanbul. 1942. 119 + 59 pp.

## Chapter II

1. DAVIDSON, J., 1936: *Trans. Roy. Soc. South Australia.* **60,** *88.*
2. KINCER, J. B., 1928: *Mon. Weather Rev.* **56,** *301.*
3. BODENHEIMER, F. S., 1926: *Z. angew. Ent.* **12,** *91.*
4. ZWÖLFER, W., 1934: *Z. angew. Ent.* **21,** *333.*
5. BODENHEIMER, F. S., 1927: *Bull. Soc. ent. Fr. 195.*
6. BODENHEIMER, F. S., 1927: *Z. angew. Ent.* **13,** *473.*
7. KLEIN, H. Z., 1930: *Anz. Schädlingsk.* **6,** *97.*
8. BODENHEIMER, F. S., 1925: *Bull. Soc. ent. Egypte, 149.*
9. ZWÖLFER, W., 1934: *Verh. dtsch. Ges. angew. Ent., Erlangen, 20.*
10. BODENHEIMER, F. S. & KLEIN, H. Z., 1930: *Z. vergl. Physiol.* **11,** *345.*
11. WADLEY, E. M., 1936: *J. agric. Res.* **53,** *259.*
12. PRADHAN, S., 1946: *Proc. Nat. Inst. Sci. India* **12,** *301.*
13. KLEIN, H. Z., 1936: *J. A. Agric. Res. Sta., Rehobot. Bull.* **21.**
14. BODENHEIMER, F. S., Citrus Entomology in the Middle East. Den Haag. 1951.
15. BODENHEIMER, F. S., Die Schädlingsfauna Palästinas. Berlin. 1930.
16. BODENHEIMER, F. S. & NEUMARK, S., The Israel Pine Matsucoccus. Jerusalem. 1955.
17. BODENHEIMER, F. S., 1943: *Bull. Soc. Fouad Ier Ent. Egypte,* **27,** *1.*
18. HECHT, O., 1936: *Bull. Soc. Ent. Egypte 299.*
19. BODENHEIMER, F. S. & GUTTFELD, M., 1929: *Z. angew. Ent.* **15,** *67.*
20. HALL, W. J., 1926: *Min. of Agric. Cairo, Techn. Bull.* **70.**
21. BODENHEIMER, F. S., 1932: *Z. angew. Ent.* **19,** *514.*
22. COOK, W. C., 1924: *Ecology, 60;* 1926: *ibidem* **7,** *376; Bull. Montana Agr. Exper. Sta. No. 225.*
23. SHELFORD, V. E., 1920: *Trans. Illinois Ac. Sci.* **13,** *257;* 1952: *ibidem* **45,** *155.*
24. SHELFORD V. E. & YEATTER, R. E., 1955: *J. Wildlife Managem.* **10,** *233.*
25. SHELFORD, V. E., 1954: *J. Mammal.* **35,** *533.*
26. WALLGREN, H., 1954: *Acta Soc. Fenn.* **84,** *1.*
27. WELLINGTON, W. G., 1951: *Can. J. Zool.* **29,** *339;* 1952: *ibidem,* **30,** *114.*
28. Part of the chapter: Physical Ecology of *Ceratitis capitata* in: Citrus Entomology of the Middle East. Den Haag. 1951.
29. BODENHEIMER, F. S., C.R.Ve Congr. Int. Ent., Paris. 1933 p. *93.*
30. BODENHEIMER, F. S., 1932: *Biol. Zbl.* **52,** *598.*
31. BODENHEIMER, F. S., Studies on the Ecology and Control of the Moroccan Locust in Iraq. I. Baghdad. 1944. p. *1-121.*
32. KNOLL, F., Insekten und Blumen. 2. Wien. 1922.
33. WEBER, H., 1931: *Z. Morph. Ökol. Tiere* **23,** *575.*
34. KENNEDY, C. H. 1925: *Biol. Bull. Wood's Hole, 390.*
35. UEXKÜLL, J. VON, Theoretische Biologie. Berlin. 1920: Umwelt und Innenwelt der Tiere. Berlin. 2nd. ed. 1921.
36. HOWARD, E., A Waterhen's Worlds. Cambridge. 1940.
37. ULLYETT, G. C., 1936: *Proc. Roy. Soc. B.* **120,** *253.*
38. ALLEE, W. C., Animal Aggregations. Chicago. 1931.

## Chapter III

1. PEARL, R., The Biology of Population Growth. New York 1926; The Rate of Living. London. 1928; 1927: *Quart. Rev. Biol.* **2,** *532.*
   PEARL, R. & PARKER, S. L., 1922: *Proc. nat. Acad. Sci., Wash.* **8,** *212.*
2. ALPATOV, W. W., 1932: *J. exp. Zool.* **63,** *85.*
3. STANLEY, J., 1932: *Canad. J. Res.* **6,** *632.*
4. GAUSE, G. F. & ALPATOV, W. W., 1931: *Biol. Zbl.* **51,** *1.*

5. ROBERTSON, F. W. & SANG, J. H., 1944: *Proc. Roy. Soc. Lond. B.* **132,** *258;* SANG, J. H., 1949: *Physiol. Zool.* **22,** *183;* 1950: *Biol. Rev.* **25,** *188.*
6. BODENHEIMER, F. S., 1953: *Science News* **30,** *44.*
7. BAR-ZEEB, M., 1957: *Bull. Res. Counc. Israel,* **6B,** *220.*
8. CHIANG, H. C. & HODSON, A. C., 1950: *Ecol. Monogr.* **20,** *173.*
9. BODENHEIMER, F. S., 1934: *Genus,* **1,** *43.*
9a. ALLEE, W. C., 1939: *Ecology,* **20,** *98.*
10. CHAPMAN, R. N., 1928: *Ecology,* **9,** *111.*
    CHAPMAN, R. N. & BAIRD, L., 1934: *J. exp. Zool.* **68,** *293.*
    HOLDAWAY, G. F., 1932: *Ecol. Monogr.* **2,** *261.*
    PARK, T., 1932: *Ecology,* **13,** *172;* 1934: *Quart. Rev. Biol.* **9,** *36.*
    STANLEY, J., 1932: *Canad. J. Res.* **6,** *632;* 1932: *ibidem,* **7,** *426.*
11. BODENHEIMER, F. S. & BEN-NERYA, A., 1937: *Ann. appl. Biol.* **14,** *385.*
    BODENHEIMER, F. S., 1937: *Quart. Rev. Biol.* **12,** *406.*
12. MACLAGAN, D. S., 1932: *Proc. Roy. Soc. Lond. B.* **111,** *432.*
13. BODENHEIMER, F. S. & SWIRSKI, S., The Aphidoidea of the Middle East. Jerusalem. 1957.
14. SWIRSKI, E., 1954: *Bull. ent. Res.* **45,** *623.*
15. BODENHEIMER, F. S., 1947: *Bull. Agric. Exper. Sta., Rehobot,* **41,** 20 pp.
16. VOSSELER, J., 1904: *Ber. Land- u. Forstwirtsch. Dtsch. Ostafrikas* **2,** *291.*
17. UVAROV, B. P., 1921: *Bull. ent. Res.* **12,** *135;* Locusts and Grasshoppers. London 1928.
18. BODENHEIMER, F. S., Studien zur Epidemiologie, Ökologie und Physiologie der afrikanischen Wanderheuschrecke. Berlin. 1930.
19. BODENHEIMER, F. S., 1932: *Biol. Zbl.* **52,** *598.*
20. KENNEDY, J. S., 1939: *Trans. Roy. ent. Soc. Lond.* **89,** *385.*
21. RAINEY, R. C. & WALOFF, Z., 1948: *J. Anim. Ecol.* **17,** *101.*
    R. C. RAINEY, 1951: *Nature, Lond.* **168,** *1057.*
22. UVAROV, B. P., Human and Animal Ecology. Unesco. Paris. 1957. p. 164.
23. BODENHEIMER, F. S., 1933: *Folia Med. Int. Orient.* **1,** *135.*
24. CELLI, A., Die Malaria und ihre Bedeutung für die Geschichte Roms und der Römischen Campagna. Leipzig. 1925.
25. KLIGLER, I. J., The epidemiology and control of malaria in Palestine. Chicago. 1930.
26. BUXTON, P. A., 1924: *Bull. ent. Res.* **24,** *289.*
27. MARTINI, E., Beiträge zur medizinischen Entomologie und zur Malaria Epidemiologie des unteren Wolgagebietes. Hamburg. 1928.
28. ROSS, R., The Prevention of Malaria. London. 1923.
29. SELLA, S., 1944: *Riv. Int. Sanita Pubbl.* Roma.
30. HARTING, E., 1893: *Zoologist* (3) **17,** *139.*
31. ELTON, C., Voles, Mice and Lemmings. Oxford. 1942.
32. SHELFORD, D. E., 1951: *Ecol. Monogr.* **21,** *149.*
33. SIIVONEN, L., *Papers on Game Res.,* Helsinki. 1957.
34. BODENHEIMER, F. S., Problems of Vole Populations in the Middle East. Res. Counc. of Israel. Jerusalem. 1949.
35. HAMILTON, W. J., 1937: *J. agric. Res.* **54,** *779.*
36. BODENHEIMER, F. S. & SULMAN F., 1946: *Ecology,* **27,** *255.*
37. FRIEDMAN, M. H. & G. S., 1934: *Proc. Soc. exp. Biol. N.Y.* **31,** *842.*
    FRIEDMAN, M. H. 1939: *Amer. J. Physiol.* **125,** *486.*
    FRIEDMAN, M. H. & MITCHEL, J. W., 1941: *Endocrinology,* **29,** *172.*
38. BROADBURY, J. T., 1944: *Amer. J. Physiol.,* **142,** *487.*
39. LESLIE, P. H. & RANSOM, R. M., 1940: *J. Anim. Ecol.* **9,** *27.*
40. BODENHEIMER, F. S. & DVORETZKI, A., 1952: *Bull. Res. Counc. Israel,* **1,** *62.* also in: 1957: *Studies in Biol. and its Hist.* Jerusalem **I,** *41.*
41. CHITTY, D., 1952: *Philos. Trans. B.* **236,** *505.*
42. FRANK, F., 1953: *Zool. Jb. Syst.* **81,** *610.*
43. BODENHEIMER, F. S., 1957: *Studies in Biol. and its Hist.* Jerusalem. **1,** *24.*

43a. FRANK, F., 1957: *J. Wildlife Management.* **21**, *113.*
43b. BODENHEIMER, F. S. & VERMES, M. P., 1957: *Studies in Biol. and its Hist.* **1**, *106.*
44. ERRINGTON, P. L., 1956: *Science*, **124**, *304.*
45. ERRINGTON, P. L., HAMERSTROM, F. & F. N., 1940: *Agric. Exper. Sta. Iowa Res. Bull.* **277,** *763.*
46. ERRINGTON, P. L., 1946: *Quart. Rev. Biol.* **21**, *144, 221.*
47. ERRINGTON, P. L., 1943: *Agric. Exper. Sta. Iowa Res. Bull.* **320,** *797.*
48. ERRINGTON, P. L., 1953: *Amer. Nat.* **85**, *273.*
49. ERRINGTON, P. L., 1954: *Ecol. Monogr.* **24**, *377.*
50. PHILLIPS, J. C., A natural history of the ducks. I. Boston. 1922.
51. CARSON, R. L., Under the Sea Wind. New York. 1952.
52. STODDARD, H. L., The Bobwhite Quail. New York. 1931.
53. LACK, D., The Life of the Robin. London. 1943.
54. FISHER, J., The Fulmar. London. 1952.

## Chapter IV

1. HOWARD, L. O. & FISKE, W. F., 1911: *Bull. U.S. Bur. Ent.* **91.**
2. SHELFORD, V., 1927: *Bull. Illinois Nat. Hist. Surv.* **16**, *307.*
   BODENHEIMER, F. S., 1928: *Biol. Zbl.* **48**, *714*; 1930: *Z. angew. Ent.* **15**, *1*; 1930: *ibidem*, **16**, *433*; 1936: *Scientia* **59**, *137.*
3. UVAROV, B. P., 1931: *Trans. R. ent. Soc., Lond.* **79**, *1.*
4. See Chapter II, 1.
5. VOLTERRA, V., 1928: *J. Cons. Int. Explor. Mer.* **2**, *3*; Leçons sur la théorie mathématique de la lutte pour la vie. Paris. 1931.
   VOLTERRA, V. & D'ANCONA, M. U., *Actual. Sci. et Industr.* No. **243**. Paris. 1935.
6. LOTKA, A. J., Elements of Physical Biology. Baltimore. 1925; 1932: *J. Wash. Acad. Sci.* **22**, *462.*
7. GAUSE, G. F., The Struggle for Existence. Baltimore. 1934; *Actual. Sci. et Industr.* No. **277**. Paris. 1935.
8. D'ANCONA, U., 1926: *Mem. R. Com. Talassograf. Ital.* **126.**
9. BODENHEIMER, F. S., 1932: *Arch. Hydrobiol.* **24**, *667.*
10. BAILEY, V. A., 1937: *Proc. Roy. Soc. Lond. A.* **143**, *75.*
11. ANDREWARTHA, H. G. & BIRCH, L. C., The Distribution and Abundance of Animals. Chicago. 1954; 1953: *Austr. J. Zool.* **1**, *174.*
12. UTIDA, S., 1953: *Ecology* **34**, *301.*
    YOSHIDA, T., 1957: *Mem. Fac. lib. Arts Miyazaki Univ.* **1**, *55.*
13. UTIDA, S., 1955: *Oyo-kontyu.* **11**, *48.*
13a. PARK, T., 1954: *Phys. Zool.* **27**, *177.*
14. NICHOLSON, A. J., *J. Anim. Ecol.* **2**. Suppl. *132.*
    NICHOLSON, A. J. & BAILEY, V. A., 1935: *Proc. zool. Soc. Lond.* **3**, *551.*
15. NICHOLSON, A. J., 1955: *Austral. J. Zool.* **2**, *9.*
16. NICHOLSON, A. J., 1954: *Austral. J. Zool.* **2**, *1.*
17. SOLOMON, M. E., 1949: *J. Anim. Ecol.* **18**, *1.*
18. RICKER, W. E., 1954: *J. Fish. Res. Bd. Canada* **2(5)**, *559.*
19. HALDANE, J. B. S., 1953: *New Biology* No. **15**, *9.*
20. HARRINGTON, W. C., 1941: *U.S. Fish and Wildlife Serv. Fish. Circ.* No. **4**, *1.*
21. VARLEY, G. C., 1947: *J. Anim. Ecol.* **16**, *139.*
22. BODENHEIMER, F. S. & NEUMARK, S., The Israel Pine Matsucoccus. Jerusalem. 1955.
23. BODENHEIMER, F. S. & SCHIFFER, M., 1952: *Acta Biotheoretica* **10**, *23.*
24. BODENHEIMER, F. S., Précis d'Ecologie Animale. Paris. 1954 p. 86.
25. LACK, D., The Natural Regulation of Animal Numbers. Oxford. 1954.
26. SMITH, H. S., 1935: *J. econ. Ent.* **28**, *873.*
27. THOMPSON, W. R., 1922: *C. R. Ac. Sci. Paris*, **175**, *651.*
28. GLOVER, P. M., 1935: *Bull. Indian Lac Res. Inst.* **22.**
29. VOUKASSOVITCH, P., 1933: *Rev. Zool. Agric.* *1.*

30. MUESEBECK, C. F. W. & DONAHAN, S. M., 1927: *Dept. Bull. U.S. Dept. Agric.* **1487**.
31. PEMBERTON, C. E. & WILLARD H. F., 1918: *J. agric. Res.* **7**, *285*.
32. FRANK, F. & ZIMMERMANN, K., 1957: *Zool. Jb. Syst.* **85**, *283*.
33. HOWARD, H. E., Territory in Bird Life. London. 1920;
    NICHOLSON, E. M., How Birds live. London, 2nd ed. 1929.
34. SACHTLEBEN, H., 1927: *Arb. biol. Reichsanst. Berlin* **15**, *437*.
35. SCHWERDTFEGER, F., Untersuchungen über die Mortalitaet der Forleule im Krisenjahr einer Epidemie. Hannover. 1935.
36. SCHWERDTFEGER, F., 1932: *Mitt. aus Forstwirt. Forstwiss.* 342.
37. SCHWERDTFEGER, F., 1936: *Mitt. aus Forstwirt. Forstwiss.* 169.
38. ELTON, C., Voles, Mice and Lemmings. Oxford. 1942.
39. BODENHEIMER, F. S., Problems of Vole Populations in the Middle East. Jerusalem. 1949.
40. BURNETT, F. M., Virus as Organism. Cambridge, Mass. 1946. p. 99
40a. ZINSSER, H., Rats, lice and History. London 1934.
41. DAVIS, D. H. S., 1953: *J. Hyg.* **51**, *427*.
42. THOMPSON, W. R., 1930: *Ann. appl. Biol.* **17**, *306*.
43. SWEETMAN, H. L., The Biological Control of Insects. Ithaca. 1936.
44. BODENHEIMER, F. S., 1932: *Z. angew. Ent.* **19**, *514*.
45. TOTHILL, J. D., TAYLOR, T. H. C. & PAINE, R. W., The Coconut Moth in Fiji. London. 1930.
46. KIRKPATRICK, T. W., The Climate and Ecoclimate of Coffee-plantations. Amani. 1935.
47. HANDSCHIN, E., 1934: *Rev. suisse Zool.* **41**, *1*.
48. HUBER, L. L., NEISWANDER, C. R. & SALTER, R. M., 1928: *Bull. Ohio Agric. Exper. Sta.* **429**.
49. THOMPSON, W. R. & PARKER, H. L., 1928: *Techn. Bull. U.S. Dept. Agric.* **59**.
50. BABCOCK, K. W. & VANCE, A. M., 1929: *Techn. Bull. U.S. Dept. Agric.* **135**.
51. ZWÖLFER, W., 1930: *Biol. Zbl.* **50**, *742*.
52. CAFFREY, 1927: U.S. Dept. Agric. Washington, 1476, 154 pp.
53. HODGSON, D. E., 1928: *Tech. Bull. U.S. Dept. Agric.* **77**.
54. BODENHEIMER, F. S., 1927: *Biol. Zbl.* **47**, *25*.
55. KOMAREK, J., 1937: *Z. angew. Ent.* **24**, *95*.
56. McATEE, W. L., 1932: *Smithson. Misc. Coll.* **85**, *7*.
57. MEHTA, D. R., 1930: *Bull. ent. Res.* **21**, *547*.
58. BODENHEIMER, F. S. & GUTTFELD, M., 1929: *Z. angew. Ent.* **15**, *67*.
59. THOMPSON, W. R., 1929: *Bull. ent. Res.* **19**, *343*.
60. BODENHEIMER, F. S., Citrus Entomology in the Middle East. Den Haag. 1951.
61. SCHWERDTFEGER, F., 1935: *Z. Forst- u. Jagdwes.* **67**, *449, 513*.

## Chapter V

1. PEUS, T., 1954: *Dtsch. ent. Z. N.F.* **1**, *271*.
2. TANSLEY, A. G., 1935: *Ecology*, **16**, *284*.
3. SOLOMON, M. E., 1949: *J. Anim. Ecol.* **18**, *1*.
4. SCHWERDTFEGER, F., 1952: *Z. angew. Ent.* **34**, *216*.
5. SCHWENKE, W., 1953: *Beitr. Ent.* **3**, *86*.
6. PETERSEN, C. J., 1913: *Rep. danish Biol. Stat.* **21**, *1*, app.; 1915: *ibidem*, **22**, *89*.
7. PETERSEN, C. J., 1915: *Rep. danish Biol. Stat.* **23**, *3*.
8. PETERSEN, C. J., 1918: *Rep. danish Biol. Stat.* **25**, *1*.
9. STEPHEN, A. C., 1933: *Trans. Roy. Soc. Edin.* **57**, *601*.
10. STEPHEN, A. C., 1934: *Trans. Roy. Soc. Edin.* **57**, *777*.
11. FORBES, E., The Natural History of the European Seas. London. 1857.
12. MOLANDER, A. R., *Kristineburge Zool. Sta.* 1877-1927 (1918) no. **2**, *1*.
13. SHELFORD, V. E., WEESE, A. O., RICE, L. A., RASMUSSEN, D. I. J. & MacLEAN, A., 1935: *Ecol. Monogr.* **5**, *249*.
14. BERG, K., 1938: *Mem. Acad. Roy. Sci. et Lettr. Danemark, Copenhague, Sci.* **(9) 8**, *255*.

15. LARSEN, E. B., 1936: *Vidensk. Medd. Naturh. Foren. Kbh.* **100**, *1*.
16. BODENHEIMER, F. S., 1935: *Arch. Naturgesch. N. S.* **4**, *88* (Orthoptera).
17. BODENHEIMER, F. S., 1932: *Bull. Soc. Ent. Egypte*, *52*; 1934: *ibidem*, *211*. (Coleoptera).
18. BODENHEIMER, F. S., Animal Life in Palestine. Jerusalem. 1935 (Other groups).
19. MENDELSSOHN H., Unpublished Studies on bird ecology and on *Leucochroa hierochuntica* in Palestine.
19a. GUINOCHET, M., Phytosociologie. Paris 1955.
20. BOYCOTT, A. E., 1934: *J. Ecol.* **22**, *1*.
21. BOYCOTT, A. E., 1936: *J. Anim. Ecol.* **5**, *116*.
22. LUNDBECK, J., 1926: *Arch. Hydrobiol. Suppl.* **17**, 473 pp.
23. KRÜMMEL, O., 1896: *Mitt. dtsch. Seefisch.-Ver.* no. **7**, 7 pp.
24. REIBISCH, J., 1911: *Wiss. Meeresunters. N. S.* **13**, *131*.
25. EHRENBAUM, E., 1909: *Arb. dtsch. Wiss. Komm. Meeresuntersuch, Helgoland. 145*.
26. JOHANNSEN, A. C., 1910: *Rapp. Cons. Explor. Mer.* **12**, no. 7.
27. EHRENBAUM, E. & STRODTMANN, S., 1904: *Wiss. Meeresunters. N. S.* **6**.
28. SHELFORD, V. E., 1932: *Bull. Illinois nat. Hist. Surv.* **19**, *487*.
29. ZWÖLFER, W., 1934: *Z. angew. Ent.* **21**, *333*.
30. BODENHEIMER, F. S., 1934: *Hadar*, **7**, *139*.
31. MAYER, A. G., 1914: *Pap. Tortugas Lab.* **6**, *3*.
32. MUNRO FOX, H., 1936: *Proc. zool. Soc. Lond. 945*.
33. MUNRO FOX, H., 1936: *Nature, Lond.* **138**, *839*; 1937: *ibidem*, **139**, *369*.
34. REICH, K., 1936: *Physiol. Zool.* **9**, *254*.
35. SHELFORD, V. E., 1932: *Ecology* **13**, *105*.
CARPENTER, J. R., 1936: *J. Ecol.* **24**, *285*.
36. BLEGVAD, H., 1914: *Rep. danish Biol. Stat.* **22**, *41*.
37. JENSEN, P. B., 1919: *Rep. danish Biol. Stat.* **26**, *1*.
38. BLEGVAD, H., 1925: *Rep. danish Biol. Stat.* **31**, *27*.
39. Unpublished experiments of the author.
40. FORBES, S. A., 1877: The Lake as a Microcosm. Bull. Peoria Sci. Assoc.
41. THIENEMANN, A., 1925: *Naturwiss. Berlin* **13**, *589*; 1926: *Verh. dtsch. zool. Ges.* **31**, *29*; 1950: *Oikos*, **2**, *149*.
42. EMERSON, A. E., 1939: *Amer. Midland Natural.* **21**, *201*.
43. BODENHEIMER, F. S., 1937: *Biol. Rev.* **12**, *393*; 1937: *Quart. Rev. Biol.* **12**, *406*.
44. FRIEDRICHS, K., 1927: *Naturwiss. Berlin*, **15**, *153*, *182*.
45. TUXEN, S. L., The Hot Springs, their Animal Communities and their Zoogeographical Significance. Copenhagen and Reykjavik. 1944. *173*.
46. LOTKA, A. J., Elements of Physical Biology. Baltimore. 1925. p. *4* ff.
47. ARISTOTLE, De Generatione Animalium. 760b. Oxford Transl. Oxford. 1912.
48. SINGER, C., History of Biology. New York. 1950 p. *17*.
49. HOWARD, L. O. & FISKE, W. F., 1911: *Bull. U.S. Dept. Agric. Ent.* no. **91**.
50. ERRINGTON, P. L., 1951: *Amer. Natural.* **85**, *273*.
51. PELSENEER, J., 1953: Actes VIIième Congres Int. Hist. Sci., Jerusalem.*465*.

## Chapter VI

1. WEINSTEIN, Chemistry in Medicine. 1928. p. *68*.
2. PEARL, R., The Rate of Living. London. 1928.
3. HANDSCHIN, E., 1932: *Pamphl. Coun. Sci. Industr. Res. Australia* **31**.
4. BREITENBRECHER, J. K., 1926: *Trans. R. Soc. Canada Sect.* **V**, *269*.
5. TENENBAUM, E., 1933: *Biol. Zbl.* **53**, *308*.
6. GLEMBOZKI, L., 1935: *Biol. Zh. Mosk.* **2**, *355*.
7. KÜHN, A. & HENKE, K., 1930: *Arch. EntwMech. Org.* **122**, *204*.
8. PEARL, R., The Biology of Population Growth. London. 1926.
9. GAGARIN, B., 1933: *Biol. Zh. Mosk.* **2**, *451*.
10. HERTER, K., 1935: *Verh. dtsch. zool. Ges. 31*.
11. GOWEN, J. W. & SCHOTT, R. G., 1933: *Amer. J. Hyg.* **18**, *674*.
12. SPENCER, W. P., 1935: *Amer. Natural.* **69**, *223*.

13. DELBRÜCK, M. & TIMOFEEFF-RESOWSKY, N. W., 1935: *Nature*, Lond. **137**, *358*.
14. HENKE, K., 1937: *Fortschr. Zool., n.s.* **1**, *460*.
15. KÜHN, A. & HENKE, K., 1929: *Abh. Ges. Wiss. Göttingen, Math.-Phys. Kl. N. S.* **1**, *1*.
16. ZIMMERMANN, K., 1956: *Zool. Jb. Syst. 463*.
17. BODENHEIMER, F. S., 1954: *Bull. Res. Counc. Israel* **4**, *31*.
18. KINSEY, A. C., The Origin of Higher Categories in Cynips. Bloomington. 1936.
19. DOBZHANSKY, T., Genetics and the Origin of Species. New York. 1954.
20. WRIGHT, S., in J. S. HUXLEY, The New Systematics. Oxford. 1942 p. *161*.
21. FORD, E. B., Mendelism and Evolution. London. 1931.
22. ELTON, C., Ecology and Evolution. Oxford.
23. DOUGLAS, J. R., 1928: *J. econ. Ent.* **21**, *203*.
24. FOA, A., 1928: *Riv. Fis. Mat. Sci. Nat., Napoli.* **2**, *1*.
25. COUSIN, G., 1932: *Bull. Biol., Suppl.* **15**, *1*.
26. ROUBAUD, E., 1926: *Ann. Inst. Pasteur, Paris.* **37**, *627*.
27. BODENHEIMER, F. S., Animal Life in Palestine. Jerusalem. 1935.
28. GOLDSCHMIDT, R., 1932: *Arch. EntwMech. Org.* **126**, *674*.
29. BODINE, J. H., 1932: *Physiol. Zool.* **5**, *538*.
30. DAWSON, R. W., 1931: *J. exp. Zool.* **59**, *87*.
31. BACOT, A., 1914: *J. Hyg. Camb. Plague Suppl.* **3**, *447*.
32. HELLER, J., 1931: *Biol. Zbl.* **51**, *259*.
33. BODENHEIMER, F. S. & VERMES, P. M., Studies in Biology and its History **1**, *106*.
34. BODENHEIMER, F. S., 1953: *Trans. 9th Int. Congr. Ent. Amsterdam.* **I**, *21*.
35. BODENHEIMER, F. S. & SHULOV, A., 1951: *Bull. Res. Counc. Israel* **1**, *60*.
36. MARCHAL, P., 1936: *Ann. Epiphyt. Phytogénét.* **2**, *447*.
37. FLOWER, S. S., 1932: *Proc. zool. Soc. Lond. 369*.
38. PATTERSON, J. T., 1927: *Quart. Rev. Biol.* **2**, *399*.
39. MARCHAL, P., 1904: *Arch. Zool. exp. gén. 257*.
40. SILVESTRI, F., 1906, 1907: *Boll. Lab. Zool. Portici,* **L** and **3**.
41. SPAERK, R., 1929: *Rep. Danish Biol. Stat.* **30**, *1*.
    ORTON, J. H., 1927: *J. Mar. biol. Ass. U. K.* **14**, *967*.
42. HUGHES-SCHRADER, S., 1926: *Science* **63**; 1930: *Ann. ent. Soc. Amer.* **23**.
43. STEINBACH, E., 1913: *Zbl. Physiol. B.* **27**, *717*.
44. Experiments performed in The Zool. Inst., Chicago.
45. WITSCHI, E., 1921: *Arch. EntwMech. Org.* **49**, *316*.
46. BALTZER, F., 1928: *Verh. dtsch. zool. Ges.* **32**, *273*; 1928: *Rev. suiss. Zool.* **35**, *225*.
47. SEILER, J., 1927: *Naturwiss. Berlin*, **15**, *33*.
48. BANTA, A. M. & BROWN, L. A., 1924: *Proc. Soc. exp. Biol. N.Y.* **22**, *77*.
49. CAULLERY, M. & COMAS, M., 1928: *C. R. Acad. Sci. Paris* **186**, *646*.
50. COBB, N. A., STEINER, G. & CHRISTIE, J. R., 1923: *J. agric. Res.* **23**, *921*
    CHRISTIE, J. R., 1929: *J. exp. Zool.* **53**, *59*.
51. LUNTZ, A., 1931: *Naturwiss. Berlin*, **19**, *585*.
52. WESENBERG-LUND, C., 1930: *K. Danske Vidensk. Selk. Skr. Sci* **(9) 2**, no. 1, *1*.
53. BERG, K., 1934: *Biol. Rev.* **9**, *139*.
54. BODENHEIMER, F. S. & SWIRSKI, E., The Aphidoidea of the Middle East. Jerusalem. 1957.
55. BRIDGES, C. B., 1925: *Amer. Natural.* **59**.
56. GOLDSCHMIDT, R., Physiologische Theorie der Vererbung. Berlin. 1927.
57. RIDDLE, O., Sex and Internal Secretions. 1932. p. *246*; 1931: *Physiol. Rev.* **2**, *63*.
58. BODENHEIMER, F. S., 1926: *Zion. Org. Agric. Exper. Sta.* no. 15. 8pp.
59. RENSCH, B., Das Prinzip Geographischer Rassenkreise. Berlin. 1929.
60. ELLER, K., 1936: *Abh. bayer. Akad. Wiss.* **36**.
61. GOLDSCHMIDT, R., 1936: *Lymantria. Bibl. Genet.* **11**, *1*.
62. BERGMANN, A., 1847: Göttinger Studien, 1. Abth. *595*.
63. HESSE, R., Tiergeographie auf ökologischer Grundlage. Jena. 1924.
64. ALLEN, J. A., 1877: *Radical Rev.* **1**, *108*.
65. ALPATOV, W. W., 1929: *Quart. Rev. Biol.* **4**, *1*.

66. GLOGER, G. L. Das Abaendern der Vögel durch Einfluss des Klimas. Breslau. 1833.
67. GOERNITZ, K., 1923: *J. Orn.* **71**, *456.*
    RENSCH, B., 1925: *J. Orn.* **73**, *127.*
68. HINTON, M. A. C., 1927: *Eugen. Rev.* **19**, *109.*
69. SUMNER, F. B., 1909: *J. exp. Zool.* **7**, *97.*
70. STANDFUSS, M., Handbuch der Palaearktischen Grossschmetterlinge. 1896.
71. JOLLOS, V., 1924: *Arch. Protistenk.* **49**, *307.*
72. BODENHEIMER, F. S. & MOSCONA, A., 1956: *Bull. Res. Counc. Israel* **5B**, *203.*
73. GOLDSCHMIDT, R., 1934: *Amer. Natural.* **68**, *5.*
74. BRUES, C. T., 1924: *Amer. Natural.* **58**, *127.*
75. HEIM DE BALSAC, H., 1936: *Bull. Biol. France et Belg., Suppl.* **21**, *1.*
76. Résultats du Voyage du "S. H. Belgica" en 1897-1899. Anvers.
77. DEWITZ, J., 1919: *Zool. Jb.* **37**, *305*; 1924: *ibidem*, **41**, *245.*
78. CLEMENTS, F. E., 1929: *J. Ecol.* **17**, *336.*
79. BOEKER, H., 1930: *Morph. Jb.* **65**, *229.*
80. DOR, M., La Morphologie de la Queue des mammifères dans ses rapports avec la locomotion. Paris. 1937.
81. RABAUD, E., L'Adaptation et l'Evolution. Paris. 1922.
82. THORPE, W. H., 1930: *Biol. Rev.* **5**, *177.*
83. STEINER, G., 1925: *Phytopathology*, **15**, *499.*
84. BRUES, C. T., 1920: *Amer. Natural.* **54**, *313.*
85. GLENDENNING, R., 1929: *Proc. ent. Soc. Brit. Columbia*, *34.*
86. HARRISON, J. W. H., 1921: *Proc. Roy. Soc. B.* **101**, *115.*
87. CRAIGHEAD, F. C., 1921: *J. agric. Res.* **22**, *189.*
88. THORPE, W. H., 1929: *J. Linn. Soc. (Zool.)* **36**, *621.*
89. DALLINGER, W. H., 1887: *J. R. micr. Soc.* **10**, *185.*
90. HINDLE, E., 1932: *J. R. micr. Soc.* **52**, *123.*
91. MAYER, A. G., 1914: *Pap. Tortugas Lab.* **6**, *3.*
92. CUÉNOT, L., La Genèse des Espèces Animales. Paris. 1911.
93. WOLTERECK, R., 1934: *Z. ind. Abst.- Vererb.-lehre* **67**, *173.*
94. WOOD JONES, F., Habit and Heritage. London. 1943.
95. THOMPSON, W. R. & PARKER, H. L., 1928: *Bull. ent. Res.* **18**, *359.*
96. CAULLERY, M., Le problème de l'Evolution. Paris 1931.

## Chapter VII

1. ZINSSER, H., Rats, Lice and History. London 1934.
2. BURNETT, F. M., Virus as Organism. Cambridge, Mass. 1946.
3. WELLINGTON, W. G., 1952: *Canad. J. Zool.* **30**, *114*; 1951: *ibidem*, **29**, *339.*
4. COOK, R. C., Human Fertility. London. 1951.
5. Great Britain: Roy. Commiss. on Population Rpt. London. 1949.
6. BERLIN, I., The Hedgehog and the Fox. Mentor Book. 1957. p. *109.*
7. HJORT, J., JAHN, G. & OTTESTED, P., The optimum Catch, in: Hvalradets Skrifter No. 7. Essay on Population. Oslo. 1933. p. *92.*
8. CARR-SAUNDERS, A., World Population: Past Growth and Present Trends. Oxford. 1936.
9. FERENZI, I., The Synthetic Optimum of Population. Paris. 1938.
10. MALTHUS, T. R., An Essay on the Principle of Population. London. 1799.
11. COOK R. C., Human Fertility. London. 1951. p. *15* f.
12. WINFIELD, G., China: The Land and the People. Rev. ed. New York. 1950.
13. COOK, R. C., Human Fertility, 1951. p. *93.*
14. BODENHEIMER, F. S., History of Biology. London. 1958.
15. RATZEL, F., Der Lebensraum. Tübingen. 1901.
16. HUNTINGTON, E., Civilisation and Climate. 3rd ed. New Haven. 1924.
17. SEMPLE, E. C., Influence of Geographic Environment, New York, 1911.
    SEMPLE, E. C., The geography of the Mediterranean region, London, 1932.
18. TAYLOR, G., Environment, Race and Migration. Chicago. 1932.

19. AYRES, Q. C., Soil Erosion and its Control. New York and London. 1936.
20. TAYLOR, F. H., The Destruction of the Soil in Palestine. Palest. Soil Cons. Board Bull. 2.
21. HEIM, H., Un Naturaliste autour du Monde, Paris. 1954.
22. PHILIPS, in: "Biology of Hot and Cold Deserts". London. 1954.
23. BATE, D. M. A., in: The Stone Age of Mount Carmel. I. Oxford. 1937. p. *141*.
24. HAGEDORN, A. L., Animal Breeding. London. 1944.
25. REMY, M., Nous avons brûlé la terre. Paris. 1954.
26. Some parts with the kind permission of the Committee for a History of Mankind (U.N.E.S.C.O.) from a manuscript submitted to it by the author.

# NOTES

note a) to p. 12. Excepted are, of course, to some degree individuals from the same egg, such as identical twins in man, or the 6 to 9 individuals born from one egg in *Tatusia*, etc.

note b) to p. 92. Just as these lines are in press some locust workers have begun to work already on a new line of thought. They agree that no fixed permanent breeding grounds exist. But they point out in detail that the weather and the ecotones of vegetation along both margins of the Red Sea show such variability that almost always an opportunity is found which is suitable for increase, aggregation and gregarization of locust swarms. Thus no fixed localities, but a definite region is regarded by them tentatively as a permanent centre of the Desert Locust. Future will show, if this Red Sea Region is really of vital permanent epidemiological importance in locust outbreaks, or if it is just a region where smaller or larger outbreak centres of the locust may occasionally appear.

note c) to p. 127. In the mean time A. J. NICHOLSON (*Ann. Rev. Entom.* **3**. 1958 p. *107—136*) has given a comprehensive review on dynamics of insect populations. We quote the end of this authoritative statement: "There is voluminous field evidence of density-induced environmental reaction which opposes population growth with increasing intensity as populations increase in size, and there are many individual examples, both in the field and in the laboratory, of populations which undoubtedly adjust themselves to their environments by inducing such reaction. In addition, logical deduction from certain well established facts has shown that density-induced governing reaction must necessarily play this role of adjustment in each persistent population."

We have always accepted this conclusion; we also accepted under NICHOLSON's influence the theory that every animal population is always in a state of "balance" with its environmental capacities. But we have robbed this ambigous term of its density-dependent interpretation. The animal population as long as it is on the lower level is not dependent upon a density as the regulatory factor. This partial acceptance of the mathematical self regulatory theory on its high population level (see fig. 21B) is all with which we can agree and we remain — we are sorry to confess it — for the lower part of the cycle in a "chaos", as we have so far no general interpretation for its trends there. In animal populations of lake and sea (RICKER), and, if we accept the general concept of adaptation of ERRINGTON, in mammals and birds, we see a first hope of interpretations which are not those of a biomathematical "cycle". In insects, inspite of VARLEY's analysis of *Urophora* (compare the recent critic of MILNE), we are still far from any such interpretation.

note d) to p. 138. While reading the proofs, we saw a study by A. MILNE (*Canad. Entom.* 1957 p. *193*) on "Natural Control of Insect Populations", with which we largely agree, even if he attacks where synthesis is desirable. His main contribution is a thorough analysis of the results of VARLEY and of the rather primitive methods still applied by the Riverside School. He shows that in the 1935/36 generation of *Urophora* the cause of death of 40 % is unknown, thus nullifying the claim that the density dependence-mechanism is fully adequate. A further detailed analysis of the mathematical procedure shows that the decisive mathematical symbols are neither precisely defined nor always fixed in an unequivocal way. Further no sampling errors are given. If these had been applied the "reliability" of the calculations would have been reduced beyond any bounds of proof. MILNE's criticisms are doubtless justified, he has, however, not proved, that VARLEY's conclusions are wrong. They do anyhow show what a high degree of qualifications is needed in order to enable a complete population analysis to be carried out. We think it obvious that scarcely any individual wil be able to perform such analysis, taking all factors into consideration. Only well integrated team work under able guidance can in future provide us with the data needed for an acceptable analysis of a natural population.

note e) to p. 143. This is in accordance with the third law of VOLTERRA.

note f) to p. 182. Generally the young generation of botanists seems to share the opinion that classical plant sociology has exhausted its potentialities and must be replaced by new methods and problems.

note g) to p. 236. A full discussion on inheritance of acquired characters took place recently in the Linnaean Society of London (*Proceed*. **169**. 1958 p. *49* ff). The two speakers on metazoans remained on the basis of orthodox Mendelism. MELVILLE proposed that reactions to the environment are first "attached" to some transmissible particle, the plasmagene, and later become incorporated into the nucleus and its chromosomal apparatus. WADDINGTON also speaks of a "genetic assimilation' as identical to the acquirement of characters. He says clearly that "the capacity of an animal to respond to a given environmental stress by a particular type of development is a function if its genotype." The details of his theoretical mechanism remain "perfectly orthodox Mendelism', even if characters, which at an early stage of evolution are produced as responses to environmental stresses, may no longer require any external stimulus at a later stage. Selection within epigenetic processes in a macro-evolutionary time scale will tend to retain these "epigenetical interactions which offer the possibility of adaptive responses to stresses."

The discussion only shows that, as in many other ecological problems, our age tends to minimize the basic contradiction in such problems which still 50 years ago seemed to be entirely incompatible (See BODENHEIMER, *Studies in Biology*. **1**. 1957). But the proposed theories actually do not help more than older ones, that in macrospeciation time (please pay attention to this term) phenocopies replace similar looking ecophenes, e.g. in geographical variation, where long ears and tails appear in the southern forms of hares and other mammals (R. GOLDSCHMIDT).

# APPENDIX

Only at the end of the proof-reading we read the admirable book of WADDINGTON "The Strategy of the Genes" (1957), illustrating his views on epigenetics, population genetics and the problems of adaptability and adaptation. The book however does not show (as all others existing do) what is, but what could be; it illustrates again the deplorable lagging behind of ecology in comparison with other questions of modern biology. We agree with WADDINGTON, that the manifold series of events in his 'epigenetic landscape' could occur and probably do occur more or less often. He has not however Lamarckism as one possible solution of his 'pseudo-exogenous' influences. His interpretation as a selection of pre-existing genes in adaptation may be true, e.g. in many cases of BERGMANN's or ALLEN's rules, but that has still to be proven. The theoretical contrast between neo-darwinism and neo-lamarckism has lost much of its sharpness. This agrees well with general tendency of biological antitheses in the presence.

The most important conclusion of WADDINGTON's book for us is the integration between the organism and the rest of the world into one undivisible complex. Only mutations still permit not this complete integration in his model of the epigenetic landscape. This integration of organism and what has been called before 'environment' agrees entirely with the modern trend in ecology.

# INDEX

## A

abundance, 40, 48, 51, 114, 128, 129, 132, 140, 144, 147, 158, 159, 164, 181, 189, 194, 196, 210
acclimatization, 50, 154
activity, 42, 50
adaptation, 10, 84, 210, 227 ff, 230, 233
adult life-span, 19
— longevity, 69, 75
after-effect, 105
age of population, 37
— class, 103
— distribution, 35, 36, 37, 103, 120
— structure, 31, 33, 35, 76, 82, 118, 238
aggregation, 88, 89, 92, 267
agreement between form and function, 229
air-humidity, 14
ALLEN's rule, 222, 269
amplitude, 131, 135
analogy, 8, 164, 198, 203
animal communities, 10, 164 ff, 168
— populations, 67, 112
aphid populations, 83 ff
arrested development, 214
ARRHENIUS curves, 42
asociality, 103
associations, 168, 175, 182, 189, 190
autecology, 166, 190
automatic adjustments, 201
autosomes, 219
autumn remigrantes, 84
average life-expectance, 30
— longevity, 44

## B

balance, 105, 108, 112, 125, 139, 155, 156, 190, 267
balancing mechanism, 124
beginning of individual life, 13
behaviour, 56, 63 ff, 88, 92, 165, 187, 189, 192, 257
benthal communities, 168
BERGMANN's rule, 222, 269
bifactorial maps, 141
biocoen, 165
biocoenoses, 190, 198
biocoenotics, 8, 167
bio-community, 189
biogeographical regions, 184
biological control, 153 ff
— density-factor, 79
— effects on population levels, 147 ff

— equilibrium 112 ff, 156 ff, 164, 198, 199
— races, 232
bio-mass, 164
biome, 175
biotic potential, 108
biotope, 105, 166, 177, 188, 190
birth-rate, 79, 244
body temperature, 42
bonitation, 60, 63, 185
— index, 57
British Molluska form no associations, 183 ff
buffering, 201

## C

cannibalism, 133
carrying capacity, 104, 125
— — of environment, 78
catastrophes, 143
catenary curve, 41
chains of relations, 200
changes of adult physiology, 75
— of behaviour, 87
charcoal production, 249
circulation of matter, 193
climate, 40, 49, 61, 84, 91, 92, 93, 114, 125, 140, 144, 149, 154, 157, 160
climatic bonitation, 62
— control, 113 ff
— destruction, 145
— index, 58, 60
— potential, 63
climax series, 196
climogram, 51 ff, 55, 59
cline, 208
coaction, 190, 192, 197, 199
cold-storage, 14
community, 169, 174, 179, 180, 184, 189, 190, 191, 192, 193, 194, 197, 199
compensation, 105, 120, 132, 133, 134, 138, 163, 164, 167
— effects, 123
— mechanism, 103
compensatory reaction, 122, 125
— tendencies, 106
competition, 128, 133, 183, 188, 189
computation, 41
concentration, 87, 88, 89, 91
condensation potential, 104
conditioning, 134
conjugation, 32
conservation of the averages, 116

constant daily oviposition rate, 37
constitution, 202, 213, 227
constitutional potency, 156
contracting population, 31, 76
control, 164
controlling factors, 84
convergence, 89
— in geographical and phenotypical variation, 233
convergent development, 230
cooperation, 197, 236
copulatory stimulation, 100
corrected temperature-sum rule, 41
cosmic radiation, 206
chromosomes, 202, 216, 218, 220
crash, 105
crisis, 92
critical period, 53
crop rotation, 253
crowding, 107, 123
cubic parabola, 42
cycles, 98, 104, 105, 115, 118, 125, 126, 129, 141, 164
cyclic fluctuations, 98
— self-intoxication, 212
cyclonic activity, 56

D

damage bonitation, 58
dauermodifications, 224, 226, 235
Dauerwald, 249
days-degrees, 26, 44
death, 13, 26, 30, 133
— curve, 23, 25
— rate, 23, 25, 26, 30
deduction, 200, 201
deforestation, 249
density, 26, 68, 69, 93, 105, 106, 120, 124, 126, 132, 143, 144, 145, 153, 156, 162, 168, 188, 203, 228
— centres, 154
— dependent, 78, 108, 126, 127, 131, 136, 137, 138, 140, 141, 143, 144, 145, 152, 156, 164, 167, 198, 267
— — mortality, 128
— — mutation pressure, 127
— factor, 67, 78, 79
— independent, 128, 129, 131, 132, 156, 164, 167
— reactions, 141
— threshold, 127, 162
desert, 49, 56, 61, 180, 184, 227, 228, 248
— locust population, 86 ff
development, 38, 57
— mechanics, 202
— stages, 44

— threshold, 41, 42, 44, 45, 46, 48, 57, 68, 86
— zero, 45
diapause, 10, 49, 50, 83, 211 ff
dispersal, 153
distribution, 187
distributional life-chart, 44
disturbance of the averages, 116
dominant, 166, 176, 180, 189, 194, 195
dormant, 213
— period of purgation, 212
double logarithmic curve, 42
drift, 210
duration of day, 213, 214
dwindling population, 30
dysharmonic habitat, 189
dynamic community concept, 180
— ecology, 9
dynamics of population, 120
— of weather, 89

E

ecoclimogram, 47, 51 ff
eco-equal species, 119
ecological ages, 37 ff
— association, 151
— chain, 93
— change, 104
— conditions, 12
— field studies, 111
— life-expectation, 22 ff
— lifehistory, 10
— life-tables, 28
— longevity, 9, 30
— long-term planning, 256
— mortality, 14
ecology, 9, 41
— and genetics, 202 ff
econograph, 246
ecophene, 207, 268
eco-system, 165
ecotone, 91, 267
ecotype, 207
ecoworld, 9, 165, 166, 167, 187, 189
effect of losses, 110
effective temperature, 19, 46, 86
efficient temperature, 42
emigration, 151, 190, 243
empirical species combination, 188 ff
endogenous senescence, 28
environment, 9, 10, 21, 41, 92, 93, 104, 106, 109, 116, 124, 125, 164, 166, 186, 187, 189, 190, 193, 195, 202, 219, 227, 230
environmental capacity, 70
— changes, 20
— conditions, 36

— resistance 31, 82, 144
— responses, 156
— stimuli, 233
environmentalism, 244, 246
epidemics, 94, 96, 98, 102, 151, 168, 238
epidemiological average, 19, 57
— bonitation, 56 ff
epidemiology, 53, 91, 92, 221
epistemology, 199
equilateral hyperbola, 41, 42
equilibrated life-associations, 180
equilibrium, 98, 141, 160
— abundance, 135
— density, 123, 124, 125
erosion, 250
errors, 10, 11, 87
erruptions, 153
estivohibernacula, 48
evolutionary processes, 114
excess population, 109
experiment, 202, 203
exponential growth, 32, 34, 115
— increase, 33
extermination, 140, 142

**F**

faciations, 175, 194
fact of losses, 110
factor interplay, 127
farest plane, 165
fecundity, 99, 138
feminization, 216
fertile age, 100
fertility, 68, 75, 81, 87, 101, 102, 103, 122, 123, 158, 190, 205, 241
fish populations, 127 ff
fluctuating environment oscillations, 138
fluctuations, 9, 42, 82, 84, 87, 95, 115, 116, 118, 119, 128, 146, 147, 148, 155, 160, 161, 164, 167, 189, 191
food, 99, 150, 152, 153, 159, 190, 191, 192, 212
— chains, 200
— competition, 192
— density, 77
— pyramid, 190, 193, 200
— production, 240, 256
— supply, 76, 240

**G**

gamic morphes, 83
gene, 202, 208
— selection, 208
generation factor, 58, 60, 61
genetic assimilation, 268
— equilibrium, 209
— factors, 12

genetical composition, 68, 238
genetics, 19, 51, 203, 206
geographical distribution, 53
— variation, 10, 221 ff 268
geotaxis, 64
goat-grazing, 249
gonad, 53, 54
gonadotrophic substance, 100
governing mechanism, 141
gradation, 113, 149, 160, 161, 162
gradient, 177
gregarigenous centres, 91
— zone, 91
gregarious forms, 87
— phase, 88
gregarisation, 87, 88, 89, 267
groundwater level, 95
growth curve, 79
— cycle, 195

**H**

habitat, 189, 197
— concept, 184 ff
haemorrhagic disease, 110
harmony of nature, 198
heliothermic animals, 42
hereditary diapause, 212
heredity, 10, 202, 206, 212
hermaphrodites, 216
heterozygotes, 23
hibernation, 35, 47, 48, 49, 50, 86, 149, 150, 161, 211
history, 189
holocoen, 197
home, 87
homology, 8
host abundance, 58
— factor, 58, 60, 61
— selection, 65, 231
— sequence, 61
— suitability, 58
human demography, 79
— ecology, 10, 237 ff
— fertility, 257
— population, 70, 78, 238
— surrounding world, 167
— survival, 257
humidity preference, 69
hyperbola, 19, 26, 41, 42, 57, 68, 69
hysteresis, 17

**I**

immigration, 117
incubation, 54
intelligence quotient, 238
intensity of struggle for existence, 82

intuition, 200, 201
interacting populations, 117, 119
— species, 118
interaction of environment and heredity in sex-determination, 216 ff
intercompensation, 106, 108
intergration, 198
interrelation, 92
intersexes, 220
interspecific competition, 118
— regulation, 115
— struggle, 166
intertropical concentration zone, 90
— convergence zone, 89
intraspecific competition, 125, 166
— density, 108
— population pressure, 148, 149, 162
— — problems, 9
— regulation 115
— strive, 103, 107, 110
isolated populations, 210
isolation, 209
isothanate, 186

## L

lag, 164
— effects, 120, 126
— period, 124
Lamarckian theory, 10, 232, 234, 236, 269
larval density, 77
law of the minimum, 157
legislative, 141
lethal genetics, 207
life cycle, 41, 44 ff, 113, 162, 200
— expectation, 28, 29, 100, 101
— history, 9, 22, 40 ff, 42, 44, 51
— intensity, 9
— space, 218
— span, 30, 38
— tables 22, 26, 27, 28, 30, 37
limited growth, 33
local gradation, 53
logistic curve, 32, 42, 67 ff, 70, 76, 78, 79, 115, 125, 132, 240, 251
longevity, 9, 14 ff, 17, 23, 26, 27, 38, 39, 67, 68, 100, 190, 193, 195, 197, 204
— of animal communities, 195
— of field-bee, 17 ff
lunar cycles, 98

## M

maintenance, 238, 247
malaria epidemiology, 10, 92 ff, 95, 97
malign tumor, 207
mammalian embryo, 13
— tail and environment, 229, 230

man-made desert, 247, 248
marine bottom communities, 168 ff
masculinization, 216
mathematical equations, 124
— interpretation, 138
maximum, 21
— longevity, 101
mean duration of life, 24
mechanistic utilization, 254
Mendelism, 10, 268
mice-typhoid, 102
microclimate, 48, 56
migration, 46, 47, 55, 56, 64, 66, 87, 88, 89, 91, 92, 99, 104, 110
migratory power, 92
— vagility, 91
milieu fixe interne, 13, 22
mixed forests, 249
— steady-state populations, 118
mobile gypsy aphid, 84
monsoon, 89, 90
mortality, 8, 26, 28, 99, 100, 106, 122, 131, 132, 138, 139, 140, 150, 155, 190, 238, 243
— factors, 137
— rate, 28
multiple-age spawners, 128
Muskrat populations, 105
mutability, 206
mutants, 22, 24
mutation, 127, 205, 206, 233
mutual aid, 166

## N

natality, 243
natural balance, 7, 113
— control, 136, 155
— equilibrium, 247, 251
— fluctuations, 251
— selection, 125
neurotoxic substance, 100
niches, 116, 144, 166, 188
non-compensation, 108, 132, 134
non-reactive factors, 122
non-reproduction, 102
numeric abundance, 113

## O

optimal catch, 239, 240, 251, 252
— density, 239
— population density, 240
optimum, 21
organism, 9
ornithosis, 237
oscillating equilibrium, 251
— population, 252

oscillation, 124, 125, 128, 130, 131, 135, 143, 160, 161, 164
outbreak-crash type, 78, 103
outbreak dynamics, 92
outbreaks, 10, 87, 88, 89, 91, 92, 98, 102, 103, 145, 150, 267
overcrowded fly population, 123
overpopulation, 106
oviposition, 42, 74
oxidation rate, 220

## P

paired factors, 51, 56
parabola, 41
parameters, 141
parthenogenesis, 83, 84
percentage of possible sunshine, 53, 54
periodic cycle, 116
permanent solitary populations, 88
phase theory, 87
phases, 88
pheno-copies, 235, 268
philosophies of life, 256
philosophy, 7, 8, 165, 167
photoperiod, 55, 56
phototaxis, 64
physical conditions, 152
— ecology, 40, 41 ff, 69, 168, 186
— environment, 54, 66
— equilibrium, 156
— survey, 184
physiological adaptation, 186
— disturbance, 104
— fertility, 100
— life-expectation, 12 ff
— life-tables, 28
— longevity, 9, 30
— optimal conditions, 21
— processes, 21
— reconstruction, 20, 158, 187, 205
physiology, 9, 41, 51, 84, 87, 92, 193, 195
plant associations, 188
— sociology, 166
plasmogens, 226
population, 7, 8, 9, 10, 36, 37, 54, 68, 70, 72, 82, 83, 84, 88, 102, 103, 106, 115, 119, 120, 122, 123, 124, 127, 133, 143, 146, 153, 156, 160, 161, 166, 167, 185, 198, 208
— abundance, 52
— analysis, 80, 81
— census, 124
— cycle, 126
— density, 24, 76, 77, 81, 137, 212, 251
— dynamics, 10, 127, 267
— equilibria, 78
— genetics, 126, 202, 207 ff, 211

— growth 30, 31, 34, 67, 70, 71, 72, 76, 78, 82, 267
— pressure, 82, 115, 209, 219, 243
— reduction, 244
— regulation, 126, 128
— size, 209
— trend, 85, 151
potato famine, 243
potential, 105
— growth, 35, 73
— increase, 32, 82
polyphagy, 158
post-reproductive period, 38
pre-adaptation, 233, 235
predation, 101, 105, 110, 111, 112, 132, 133
predator-prey relationship, 106
predominants, 166
preferendum, 177
pre-inductions, 235
pressure, 190
preventive medicine, 238
principle of compensation, 8
prolan, 100
psittacosis, 237
psychology, 164

## R

rain, 237
— factor, 60, 61
— fall, 92
rate of egg laying, 76
— of metabolism, 17
reaction basis, 188, 190, 212, 227
real adaptation, 233
recruitment, 133
recurrent mutations, 209
— weather catastrophes, 142
reflex, 64
regulating factor, 138
regulation, 126, 140, 141, 153, 156, 164
regulatory mechanism, 104
reliability of rainfall, 89
reproducing population, 9
reproduction, 20, 32, 38, 44, 100, 102, 104, 131, 132, 133, 238, 242
— curves, 131, 134
reproductive period, 38
— potential, 83, 107
reshuffling of gene combination, 210
restricted mortality, 239
reversible processes, 209
Romanticism, 165

## S

savannah, 87

scale of activity, 70
scientific method, 200, 201
seasonal fluctuation, 53
— rhythm of the natural habitat, 216
selection, 189
— pressure, 111, 209
— process, 185
selfgoverning systems, 124
selfregulating populations, 126
selfregulation, 190, 198
semi-estivation, 45, 46
Sense Ecology, 63 ff
sensitive period, 213
sex, 218
— determination, 10, 216 ff, 219, 220
— reversal, 219
sexuales, 84
sexuparae, 83
shock disease, 105
sigmoid curve, 67, 78
single-age spawners, 128
sociability, 104
social appetite, 66
— temperament, 188
sociology, 197
solitary forms, 87
— phase, 88, 91
speed of development, 77
— of life, 14
stable equilibrium, 114, 130
— human population, 240
— oscillation, 129
— population, 30, 31, 76, 100, 101
— — density, 124
stagnant population, 37
staple food, 83
starvation, relative at higher temperature, 20
statistical conception, 197
steady densities, 138
steppe, 61, 64, 166, 184
sterility, 103
stock-recruitment, 128
— reproduction, 130
stress, 104
strong periodic fluctuation, 210
sub-dominants, 176
success in finding food, 64
sum of effective temperatures, 48
sun radiation, 43
sunspot cycles, 98
superorganismic, 192, 197
— biocoenosis, 167
— conception, 190 ff
— structure, 180, 197
supraorganismic, 192, 193, 200
— biocoenosis, 199, 200

— integration, 10, 195
surrounding world, 165, 257
survival, 23, 24, 25, 26, 28, 30, 84, 103, 114, 209, 241
symphony of the spheres, 198
synchronism between optimal conditions and period of activity, 162
synecology, 9, 190, 199
synoptic chart, 89
synthesis, 7, 8, 11
synthetic optimum of population, 240

## T

taxis, 189
temperature, 14, 16, 17, 20, 21, 24, 45, 46, 49, 50, 51, 55, 56, 57, 64, 67, 68, 94, 96, 98, 99, 112, 113, 132, 192, 195, 206, 224, 233
— dependency, 41, 42
— preference, 69
— sum rule, 19
— threshold, 47, 211
terrestrial associations, 176 ff
territories, 153, 159
thermal constant, 26, 41, 42, 57, 68, 212
— sum, 20, 42, 44
thermo-hygrogram, 186
threshold 21, 26, 96
— of tolerance, 212
toilet habits, 234
toleration, 190
— limits, 109
total egg-production, 35
— life-intensity, 14
toxicosis, 14
trade-wind, 89, 90
trend of parasitation, 85
triple band of hyperbolae, 43

## U

ultra-violet, 54
— -radiation, 53, 98
unlimited growth, 31, 32, 33
upper heat tolerance, 233
— turning point, 42
utilitarian industrialization, 256

## V

vagration, 55, 84
Van 't Hoff curves, 42
variation in reproduction, 128
verge of extermination, 102
vigour, 158
vitality, 148, 159, 204, 205
viviparae alatae, 83, 84
vole fluctuations, 98 ff

## W

warm- and cold-weather fronts, 237
weather, 40, 56, 106, 113, 125, 141, 144, 150, 156, 157
— bonitation, 59, 60
— dynamics, 92
whole, 193
winter mortality, 102
worlds, 65, 66, 165

## X

X-chromosome, 219, 220

## Y

yeast, 76

## Z

zone, 172, 173, 174, 184, 185, 187
zoogeography, 40, 44, 45, 52, 53, 189, 197, 221, 222